DETERMINATION OF THE *IN SITU* MODULUS OF DEFORMATION OF ROCK

A symposium
presented at the
Winter Meeting
AMERICAN SOCIETY FOR
TESTING AND MATERIALS
Denver, Colo., 2–7 Feb. 1969

ASTM SPECIAL TECHNICAL PUBLICATION 477

List price $19.25

AMERICAN SOCIETY
FOR TESTING AND MATERIALS
1916 Race Street, Philadelphia, Pa. 19103

Library of Congress Catalog Card Number: 71-113087
ISBN 0-8031-0061-2

NOTE

The Society is not responsible, as a body,
for the statements and opinions
advanced in this publication.

Printed in Bethlehem, Pa.

June 1970

Foreword

The Symposium on the Determination of the *In Situ* Modulus of Deformation of Rock was given at the Winter Meeting of the American Society for Testing and Materials held in Denver, Colo., 2–7 Feb. 1969. The session was sponsored by Committee D-18 on Soil and Rock for Engineering Purposes. The presentation of the papers was followed by a panel discussion: Should *In Situ* Tests for Determining the Modulus of Deformation of Rock Be Standardized? moderated by G. B. Clark and D. U. Deere. Don U. Deere, University of Ilinois, presided as symposium chairman.

Related ASTM Publications

Instruments and Apparatus for Soil and Rock Mechanics, STP 392 (1965), $8.50

Testing Techniques for Rock Mechanics, STP 402 (1966), $14.75

Determination of Stress in Rock—A State-of-the-Art Report, STP 429 (1967), $4.75

Performance of Deep Foundations, STP 444 (1969), $20.00

Contents

Introduction

Nearly every problem in rock engineering and construction involves either the strength of the *in situ* rock mass or the compressibility of the rock mass. For purposes of design it is necessary to represent in equations of engineering mechanics the corresponding numerical values representing the appropriate *in situ* property.

Strength values and modulus values determined from laboratory testing of intact rock cores are recognized as not being directly applicable to the *in situ* rock mass because of the scale effect. The *in situ* mass invariably contains joints and other types of geologic discontinuities, which may be spaced from just a few inches up to several feet apart. The engineering structure will usually stress a mass of rock sufficiently large to include several of these discontinuities. Laboratory tests, on the other hand, are performed on intact rock cores containing no discontinuities (or, at most, a single joint).

In recognition of this situation, engineers have more and more over the last two decades gone to large-scale field testing, particularly with respect to design studies for large concrete dams and related structures. With the advent of rock mechanics and its development to the status of a recognized discipline in the 1960's, there has been even greater emphasis on field testing. New methods and concepts of field testing have recently been put forward by several workers at various organizations for both *in situ* shear tests and *in situ* modulus of deformation tests. The present symposium treats the latter.

The deformation modulus is used not only in the design of concrete dams, particularly arch dams, but also in the design of steel liners in pressure tunnels, and in settlement analyses of heavy bridge piers, high-rise structures, and nuclear reactors founded on rock. Displacement calculations for the arch and walls of large underground openings also make use of the modulus.

Because of the wide interest in the subject and its timely nature, the organizing committee felt it would be appropriate to bring together specialists and organizations who are currently using the new and varied methods —not so much for the sake of attempting to standardize the procedures but as an opportunity to compare methods, results, and possible future developments and to make this information available to the engineering profession.

The interest was certainly manifested by the lively discussion from the floor and from the panel participants, R. Goodman, A. Hendron, G. Kruse,

M. Rocha, and G. Wallace on the topic: Should *In Situ* Tests for Determining the Modulus of Deformation of Rock be Standardized? moderated by G. B. Clark and myself. The general, although perhaps not unanimous, opinion was that the time was not ripe for standardization of any one of the methods at this time because of the newness of some of the methods and the additional developments to be expected in the near future. Perhaps to be more preferred would be the early compilation of recommended standards of practice for each of the tests.

Special thanks are given to the authors of the papers, to the members of the panel, to Cochairman E. J. Deklotz, to Comoderator of the panel discussion G. B. Clark, and to all members of Subcommittee 12 on Rock Mechanics of Committee D-18 for their cooperation and support.

Don U. Deere

Professor of civil engineering and of geology,
University of Illinois, Urbana, Ill.;
symposium chairman.

G. B. Wallace,[1] *E. J. Slebir,*[1] *and F. A. Anderson*[1]

In Situ Methods for Determining Deformation Modulus Used by the Bureau of Reclamation

REFERENCE: Wallace, G. B., Slebir, E. J. and Anderson, F. A., "***In Situ* Methods for Determining Deformation Modulus Used by the Bureau of Reclamation,**" *Determination of the* In Situ *Modulus of Deformation of Rock, ASTM STP 477,* American Society for Testing and Materials, 1970, pp. 3–26.

ABSTRACT: More complex foundations are being considered for concrete dams today than in past years. Large-scale *in situ* jacking tests capable of obtaining more data on rock behavior and jacking equipment which can apply total loads of 900,000 lb (408,240 kg) to 32,000,000 lb (14,515,000 kg) and measure up to 66 sets of rock deformations have been developed. Total rock surface areas under a maximum load of 1000 psi (70 kg/cm^2) varied from 6.25 ft^2 (0.58 m^2) to 225 ft^2 (21 m^2). Hydraulic rams and large flatjacks were used to apply the load to rock. The large-scale jacking tests bring fractures and joints into play and measure their effects at depth by precisely monitoring the numerous anchor displacements in NX boreholes.

KEY WORDS: rock mechanics, rocks, boreholes, joints (junctions), dams, deformation, modulus of elasticity, foundations, tests, evaluation

As the choice of ideal damsites continues to diminish, accurate predictions of the load-time-deformation property of rock foundations for concrete dams become more important. Changes in rock types, jointing and shear zone characteristics, foliation, and geological structure all contribute to variations in the deformation modulus. Such variations between the top of the canyon and the valley floor from dam-foundation contact throughout the stressed rock mass may have an important influence on the design of the structure.

[1] Engineers, Structural and Rock Mechanics Section, Division of Research, U.S. Bureau of Reclamation, Denver, Colo. Mr. Wallace and Mr. Anderson are personal members of ASTM.

The current group of Bureau foundation tests began at Glen Canyon damsite and has continued through Yellowtail and Morrow Point damsites. The largest rock mechanics test program ever undertaken by the Bureau is presently in progress at Auburn damsite about 40 miles northeast of Sacramento, Calif. During this period of development, three basic configurations of equipment have been used to measure the *in situ* deformation moduli:

1. Dual-axial jacking test (heavyweight hydraulic rams).
2. Uniaxial jacking test (lightweight aluminum columns and circular flatjacks).
3. Radial jacking test (ring sets and rectangular flatjacks).

The basic principle of dual-axial, uniaxial, and radical jacking tests is the same, that is, a load of known intensity is applied to a measured area of carefully prepared rock surface. The resulting rock deformation is then measured, and the deformation modulus is calculated by applying elasticity theory to the particular configuration of the loaded rock mass for each of the three tests. The realization that jointed rock does not behave elastically has prompted labeling the results as deformation moduli rather than elastic moduli. Current measurement systems are capable of distinguishing between joint closure and elastic deformation of intact rock within the rock mass. In time, design procedures may be expanded to completely utilize these basic components of deformation.

Test Equipment

Dual-axial Jacking Test

The equipment configuration used by the Bureau, beginning in 1961, is shown in Fig. 1. It utilizes 200-ton (181,437-kg)-capacity, double-acting hydraulic rams to apply loads to the rock. In an effort to test a large volume, two such rams are used to load two areas of rock, each 24 in. (61 cm) in diameter and 3 ft (0.91 m) apart at the centers. Upward loads from each jack are carried to the arch of the tunnel through a rigid metal shoe, timber blocking, an oak cushion, and a 24-in. (61-cm)-diameter concrete pad 3 in. (7.6 cm) thick. Downward loads from each jack are imposed on the invert of the tunnel through a rigid metal shoe, a hydraulic cushion, and a 24-in. (61-cm)-diameter concrete pad 5 in. (12.7 cm) thick.

Two fluid pressure systems, active and reactive, are incorporated in the load scheme. In the active system, which is connected to the jacks, hydraulic fluid pressures up to 10,000 psi (700 kg/cm^2) are maintained by an electric pump. The reactive system is connected to the hydraulic cushions resting on the concrete pads beneath each jack. Each cushion consisted of a flat piston 24 in. (61 cm) in diameter, with the upper end

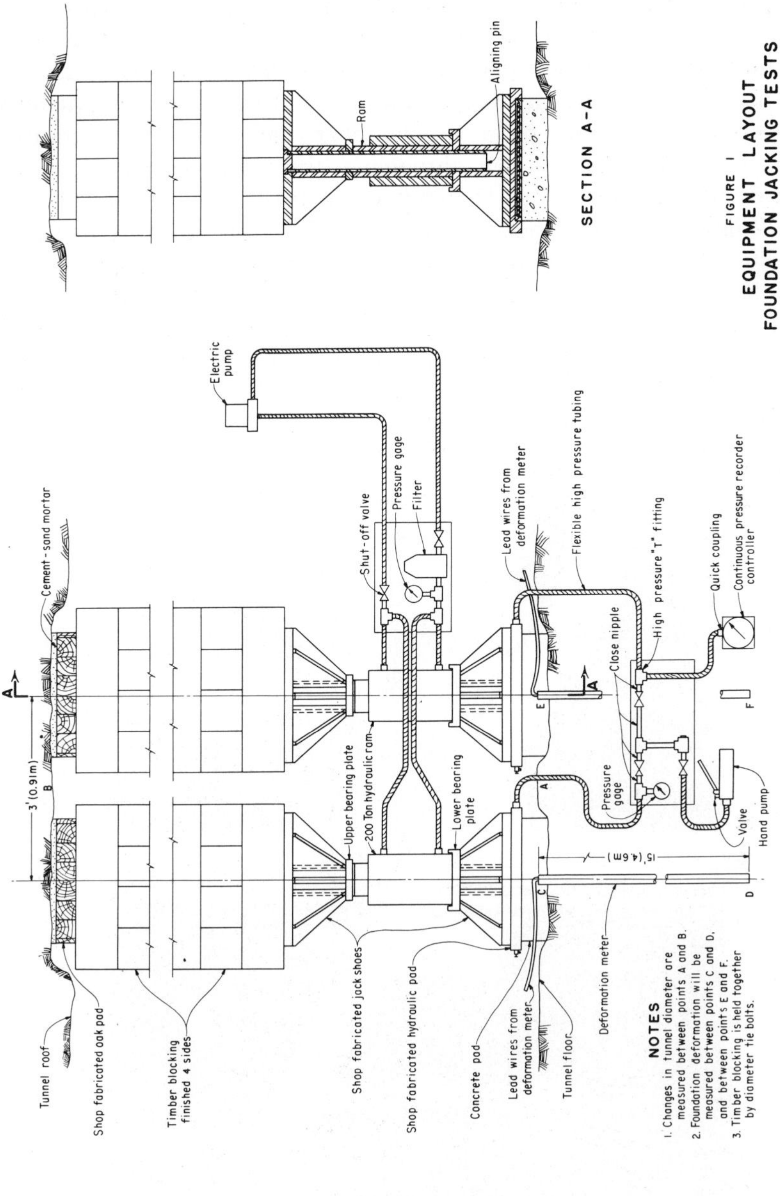

FIG. 1—*Equipment layout for dual-axial jacking test.*

inserted in a thin-fixed head cylinder and the other bearing on the concrete pad. A hand pump is used to maintain a volume of oil between the cylinder head and piston sufficient to insure uniform distribution of the load through the concrete pad onto the rock. The cushions under both jacks are connected to a hydraulic recorder-controller which continuously records the oil pressure on a circular chart. This provides a complete history of the load on the rock throughout the test, since the pressure in the hydraulic cushions is at all times equal to the load intensity on the rock. When pressure in the cushions falls off because of rock movement or slight leaks, the recorder-controller activates the pump supplying oil to the jacks until the desired pressure in the cushions is restored.

The equipment can accurately exert a maximum pressure of 800 psi (56 kg/cm^2) to the rock for indefinite periods although most of the testing

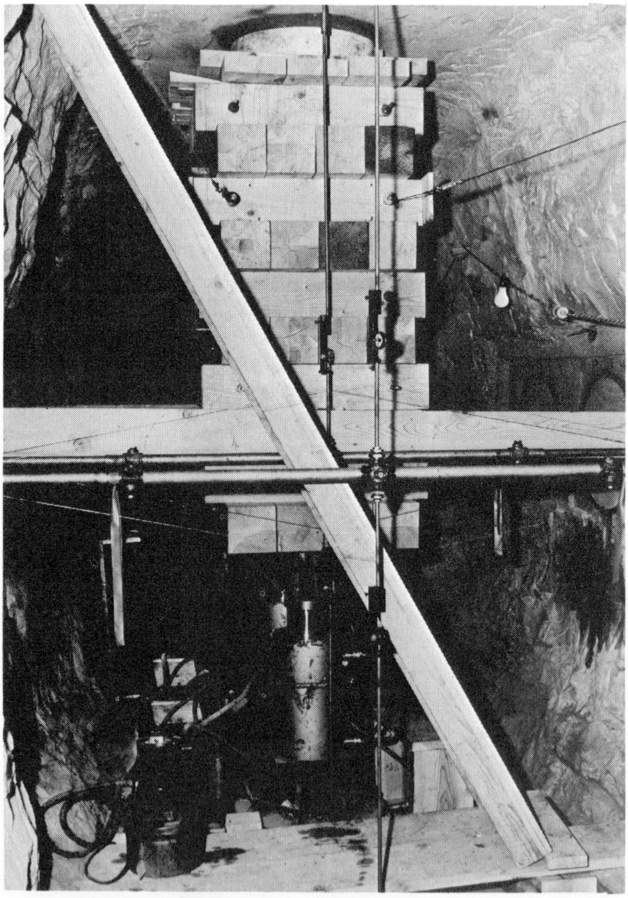

FIG. 2—*Dual-axial jacking test equipment installed for testing.*

has been limited to a maximum pressure of 600 psi (42 kg/cm^2). A typical load sequence is 200, 0, 400, 0, 600, and 0 psi (14, 0, 28, 0, 42, and 0 kg/cm^2). The nonzero increments are each held for 6 days, while the zero (recovery) increments are of 1-day duration. Figures 2 and 3 show field installations of the dual-axial jacking equipment. Vertical loads are applied simply by stacking the equipment from the invert of the test adit

FIG. 3—*Hydraulic rams used to apply loads with dual-axial test equipment. Pressure recorder-controller is at right of photo.*

up to the arch as shown in the figures. The application of the horizontal loads is done with the same equipment turned 90 deg and placed on timber staging.

Deformation of the rock midway between the two jacks is measured by a tunnel diameter gage which measures the increase in span between the arch and the invert as load is applied. It consists of a length of invar rod extending from the invert and another extending from the arch, with a dial gage between them to measure their relative movement.

In addition to the tunnel diameter gage, a special commercial joint meter is installed in an AX drill hole beneath each jack. Each joint meter measures the deformation of the rock directly underneath its companion jack to a depth of about 15 ft (4.57 m) with a pair of resistance coils wound in such a way that any change in the length of the meter causes a proportionate change in the resistance of the coils. This length-resistance change relationship is linear and repetitive, and the meter can be calibrated in fractions of an inch. The sensitivity is ± 150 μin. (0.0004 cm). The nominal length of the joint meters as purchased is 10 in. (25.4 cm); however, they are attached to 14 ft (4.21 m) of steel rod to give an overall gage length of nearly 15 ft (4.57 m). A flange at the top of the meter is cast into the concrete pad at the collar of the drill hole, and the lower end of the rod is grouted at the bottom of the hole. The rod is held centered but unattached to the wall of the hole throughout the rest of its length. A waterproof cable from the gage extends through the concrete pad to a full bridge readout terminal.

For the field test results to be of maximum benefit, the data from *in situ* deformation tests must be gathered during the field exploration phase. This means testing at remote, difficult sites without the benefit of access roads, materials handling equipment, or manpower, all of which are not generally available until the construction phase. Since the dual-axial test equipment weighed about 3 tons (2720 kg), with four individual pieces each in excess of 500 lb (227 kg), it was very difficult to transport and install under these conditions. This led to the development of lightweight unaxial test equipment, which has proven to be very efficient and effective.

Uniaxial Jacking Test

During the use of the dual-axial jacking test equipment at Morrow Point damsite, experimentation began on the suitability of using flatjacks to apply test loads to rock. These and additional studies in the laboratory indicated that flatjacks and aluminum columns could be used effectively and safely in such an application. At the same time the Bureau began developing this lightweight equipment, Shannon and Wilson Inc.[*1*],[2] Seattle, Wash., also became interested in the same objective. Ideas freely exchanged between the two organizations facilitated the development. Shannon and Wilson completed their development ahead of the Bureau and made first use of it at Dworshak Dam. The Bureau equipment, shown diagrammatically in Fig. 4, fabricated in Denver and tested in an exploratory adit about 35 miles (56 km) southwest of Denver, had the benefit of the Morrow Point and Dworshak experiences. Two components of this equipment weigh 70 lb (32 kg) each, but the remaining components each weigh less than 40 lb (18 kg). The total shipping weight is about 1000 lb (454 kg), providing a very significant advantage over the dual-axial equipment at remote sites.

A flatjack is essentially a metal membrane consisting of two metal sheets brazed together at the edges in various geometrical configurations, such as

[2] Italic numbers in brackets refer to the list of references at the end of this paper.

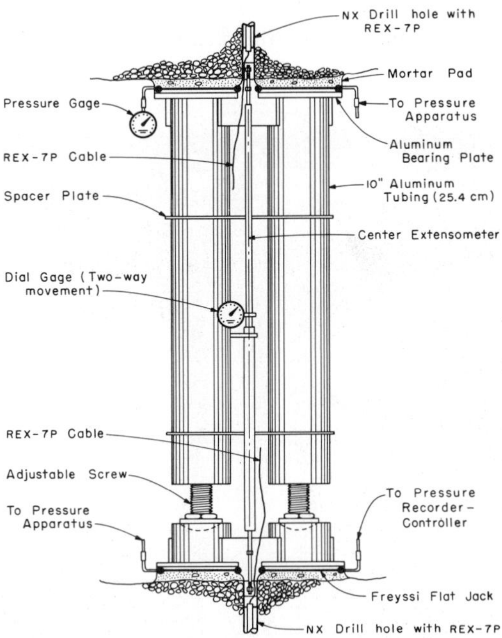

FIG. 4—*Section through uniaxial jacking test setup.*

circular, rectangular, or triangular. It is capable of holding a fluid under pressure and transmitting the pressure to a test surface. Circular flatjacks were chosen for the uniaxial test. The edges of the jacks were rounded and enlarged to permit greater extension as the test surface was deformed; even so, the expansion of each flatjack was limited to about 1 in. (2.54 cm). For competent rock, one flatjack on each end of the aluminum columns shown in Fig. 4 is sufficient to maintain the test load as the rock deforms; but in severely jointed or highly compressible rock, additional flatjacks may be required.

Oil is pumped to the flatjacks by a hydraulic system having a capacity of 1500 psi (105 kg/cm^2), although the jacks only have a maximum capacity of 2000 psi (140 kg/cm^2). The outside diameter of the flatjacks is 34.2 in. (87 cm), with a 2-in. (5.1 cm) hole in the center. The load is transmitted to the rock through a 36-in. (91-cm)-diameter concrete pad about 5 in. (12.7 cm) thick to compensate for uneveness of the rock surface. The maximum load applied to rock surfaces to date has been 1000 psi (70 kg/cm^2), which requires a slightly greater pressure in the jacks. The exact amount of pressure for any desired rock load was determined by the calibration of the jacks in the Bureau's 5,000,000-lb (2,270,000-kg) universal testing machine shown in Fig. 5. The rigidity along the enlarged edges of the flatjacks serves to make the effective area somewhat less than the true area of the jacks at lower pressures.

FIG. 5—*Calibrating uniaxial jacking test equipment in 5 million-pound Universal testing machine.*

Pressure is maintained in the flatjacks by a pressure recorder-controller similar to that used in the dual-axial tests. However, because of the direct load applications by flatjacks that also function as hydraulic cushions, only one fluid system is required. It should be noted that this system employs pressures less than 1500 psi (105 kg/cm^2), where the active system in the earlier dual-axial equipment required almost 10,000 psi (700 kg/cm^2) to achieve the specified rock load.

Figure 6 shows the equipment during a field test. The orientation of the uniaxial jacking test is completely flexible; however, the commonly used orientations are parallel to the expected line of thrust of the arch dam

FIG. 6—*Uniaxial jacking test setup parallel to resultant thrust of the dam.*

and normal and parallel to key geologic features. A typical load sequence is 200, 0, 400, 0, 600, 0, 800, 0, 1000, and 0 psi (14, 0, 28, 0, 42, 0, 56, 0, 70, and 0 kg/cm^2). The nonzero increments are each held for 2 days and the zero (recovery) increments are of 1-day duration.

Rock deformation is measured by a tunnel diameter gage and two seven-position retrievable extensometers (REX-7P). The tunnel diameter gage measured the movement between gage points directly in the center of the two opposing, loaded rock surfaces. The small holes in the center of each flatjack permitted this measurement, not possible in the earlier dual-axial tests, by eliminating the problems associated with making measurements between free surfaces outside of the areas of load application. The REX-7P measures rock strains at seven different depths beneath the loaded surfaces. They are installed in 20-ft (6.06-m) NX holes drilled to coincide with the line of action of the opposing test loads. Seven linear variable differential transformers (LVDT) are used to electronically sense the deformations between the collar of each hole and the seven expandable anchors fastened within the hole, as shown in Fig. 7. The sensing head containing the LVDT's is fixed at the collar of the hole. It also contains the gage point for the tunnel diameter gage, so both of the measurement systems are tied together by a common reference datum.

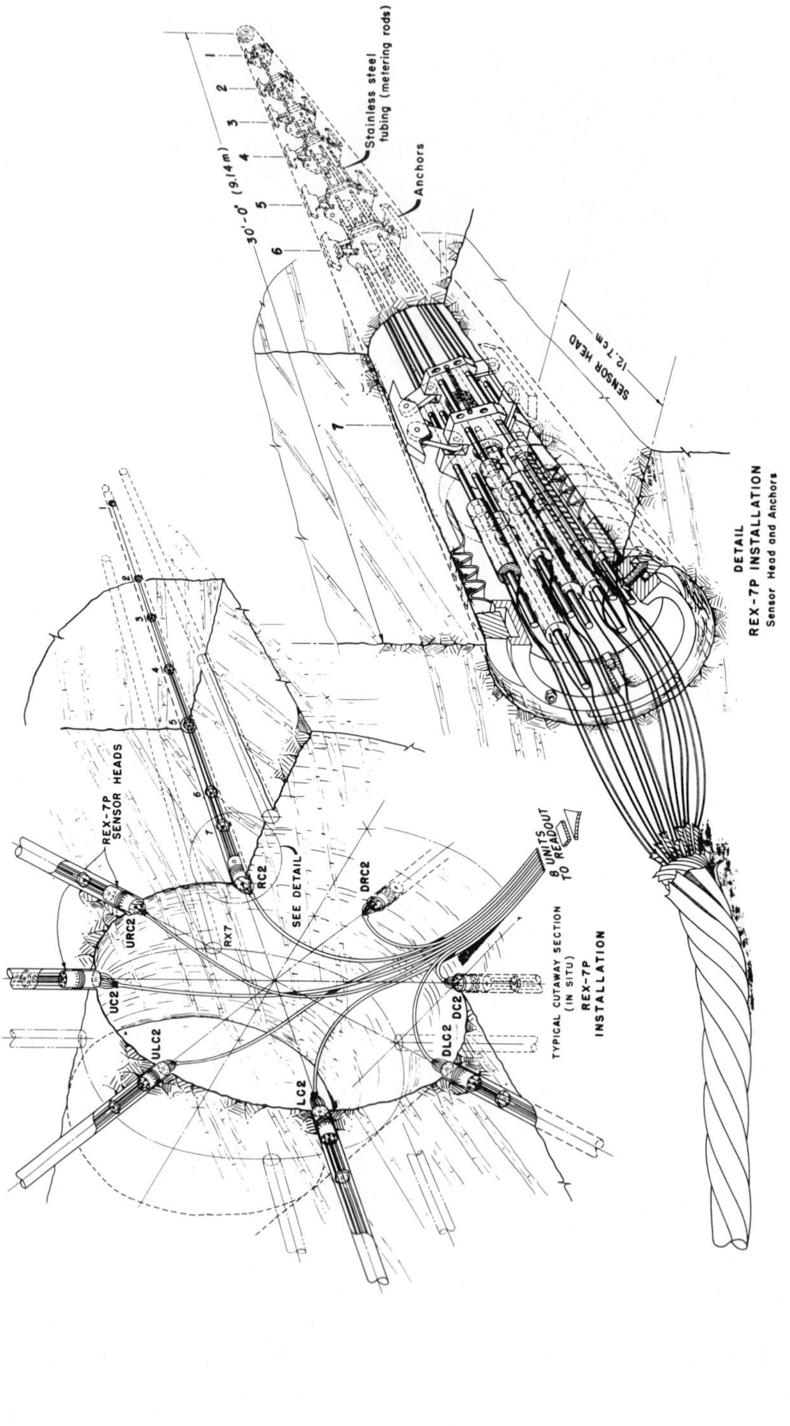

FIG. 7—*Radial jacking test, REX-7P installation.*

Metering rods of 5/16-in. (0.80-cm) stainless steel thin-wall tubing connect each of the anchors to one of the LVDT's in the sensing head. The first anchor, with its metering rod attached, is placed in the hole at the measuring point most distant from the collar of the hole, normally near the bottom of the hole. A setting rod that engages lugs on the anchor is used to position and tighten the three-legged "scissor jack" anchor. The metering rod extends from the anchor to the collar of the hole, and each subsequently placed anchor has holes through which the metering rods attached to the more distant anchors pass. After all of the anchors are positioned and tightened, the sensor head is fastened just inside the collar of the hole. Figure 8 shows two anchors and the sensor head.

FIG. 8—*REX-7 assembly. Only two of the seven anchors and the sensor head are shown.*

Each LVDT essentially consists of three encapsulated coils fixed near the collar of the hole and a movable soft iron core attached to an anchor by a metering rod (Fig. 9). A controlled ac induced in the central primary coil generates a flux that is carried by the core to the secondary coils on each side of center. If the movable core is off-center, more voltage will be induced by the flux in the closer secondary coil. The two secondary coils are connected in series opposition, and the difference in potential between them is read on a highly sensitive voltmeter. As the anchor, metering rod, and core move with the rock, the voltage change increases in the direction

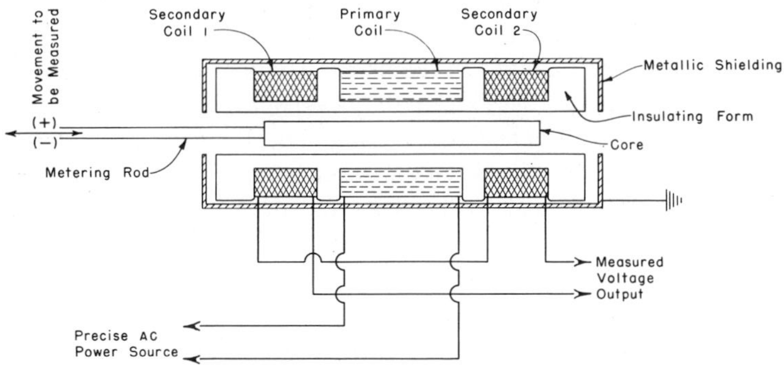

FIG. 9—*Component parts of LVDT.*

of, and proportional to, the rock movement. The REX-7P will reliably measure rock movements of 200 μin. (0.0005 cm) or more.

LVDT's have several important advantages over bonded-resistance wire gages in underground use. Bond failures and adhesive problems are eliminated since the gages are mechanically fastened to the rock. Waterproofing and lead length problems are greatly reduced. The coils of the LVDT are permanently encapsulated at assembly, and the leads are carefully waterproofed prior to installation in the REX-7P sensor head so that small changes in moisture conditions between the coils and the movable core are of little importance. However, metallic shielding of the coils should be provided and all circuitry from the power source grounded to prevent stray currents from influencing results. Without this protection voltage variations in powerlines for lights, radios, pumps, etc., can cause trouble. The chances for error due to electrical and magnetic disturbances are further reduced by using an oscillator that delivers high quality uniform voltage at 2 kHz to the LVDT. Although the uniaxial jacking tests provide a reliable measure of the deformation modulus of rock containing small, closely spaced joints and fractures, it does not fully show the influence of large, widely spaced discontinuities.

Radial Jacking Tests

In an effort to test larger areas where major discontinuities exist, the Bureau began development of a radial jacking test. Basically, the radial jacking test represents a mechanical pressure chamber. All of the equipment is retrievable and reusable, and it has an advantage over a fluid-filled pressure chamber in that fluid does not come into contact with rock. This feature permits testing of materials which would otherwise not be suitable because of leakage through joints. Nevertheless, the advantage of the pressure chamber remains, that is, being able to load a relatively large area of rock with discontinuities beyond the capability of axial tests.

The radial test equipment consists of structural ring sets with flatjacks fastened to their outer circumference and installed in a carefully excavated test bore about 8.5 ft (2.6 m) in diameter. A laboratory setup of the equipment is shown in Fig. 10. Sixteen flatjacks 8 ft (2.4 m) long by 16 in. (40.6 cm) wide are attached to the ring sets. Between each flatjack and the ring sets, a chipped wood composition board is inserted to serve as a cushion, distribute the load to the sets, and prevent damage to the jacks. As shown in Fig. 11, this board is shaped to fit the ring sets on one side and the flatjacks on the other.

Initial tests utilized 14 ring sets fabricated of steel, but these have been replaced in recent tests by 8 sets of aluminum rings. Each ring is composed of four segments bolted together, and the resulting rings are then assembled into one unit 6.66 ft (2 m) in diameter and 8 ft (2.4 m) long.

FIG. 10—*Radial jacking test equipment installed in laboratory mock tunnel.*

The rings and flatjacks are held centered in the bore of the tunnel by suspending them from the roof of the tunnel on a heavy pipe frame, and the space between the flatjacks and the rock is pumped full of concrete, as shown in Fig. 12.

Pressure is supplied to the flatjacks by a pump and hydraulic manifold. The pressure is transmitted from the jacks to the concrete and the rock surfaces. Pressure control is done manually because of the large volume of hydraulic oil needed for the jacks and to provide surveillance of this relatively complex system. To provide a uniformly distributed load, it is necessary to have enough oil in each jack to prevent the sides of the jacks from coming together during a test. The hydraulic system has proved to be very stable, with each jack showing nearly equal expansion and pressure.

Maximum pressure applied to the rock surface to date is about 1000 psi (70 kg/cm^2) although the jacks have a maximum capacity of 2000 psi (140 kg/cm^2). A typical load sequence was 200, 0, 400, 0, 600, 0, 800, 0, 1000, and 0 psi (14, 0, 28, 0, 42, 0, 56, 0, 70, and 0 kg/cm^2). The nonzero increments are each held for 2 days and the zero (recovery) increments are of 1-day duration.

The instrumentation to measure the rock deformation with load in the radial test is of three types: the REX-7P, special joint meters, and a tunnel diameter gage. The REX-7P is the primary instrument of the radial tests

FIG. 11—*Close-up of portion of radial jacking test. Area between jacks and rock to be backfilled with concrete.*

and is installed in the same manner as in the uniaxial test. Eight REX-7P's are installed in 30-ft (9.1-m) NX drill holes drilled radially outward from the center section of the test bore prior to the installation of the ring sets. These holes are drilled vertically, horizontally, and at 45 deg from vertical. Two feet (0.61 m) on each side of the central plane, four 30-ft (9.1-m) holes are drilled vertically and horizontally. These holes, and the eight in the central plane, are all drilled perpendicular to the longitudinal centerline of the test bore. The four holes on each side of the central plane are instrumented with special joint meters having a 30-ft (9.1 m) total gage length to provide additional knowledge of the rock behavior. Leads from these 16 instrumented drill holes are shielded with poly(vinyl chloride) pipe to prevent damage during the subsequent installation of the ring sets, jacks, and concrete.

FIG. 12—*Radial jacking test setup after concrete backfill is in place.*

After the equipment installation, but before placing concrete, 1-in. (2.54-cm)-diameter steel rods are inserted between the ring sets and jacks at the center section and are temporarily fastened to the set. Four rods are installed in this manner, two diametrically opposed on a nearly vertical axis and two on a nearly horizontal axis. After the concrete is placed, these rods are removed and a slightly smaller rod is installed in each hole and anchored to the rock. A stainless steel measuring plug is installed on the end of each rod to provide two orthogonal sets of diametrically opposed measuring points at the center section. A spring-loaded invar tunnel diameter gage is then used to measure the changes in tunnel diameter as the system is being loaded and unloaded.

Details of the radial jacking test equipment can be found in Ref *2*. A summary of pertinent equipment and test procedures is shown in Fig. 13.

Site Preparation

If test results are to be valid, an *in situ* test site must be carefully selected and prepared. Because of the expense of *in situ* tests for deformation moduli, they are usually confined to those foundation rock types which either predominate at the site or possess special structural characteristics. Frequently, the choosing of a "representative" rock is a difficult task. Success

FEATURES	DUAL-AXIAL JACKING TEST (Heavyweight Rams, Jointmeters, and Tunnel Diameter Gage)	UNIAXIAL JACKING TEST (A) (Aluminum Columns, Circular Flat Jacks, Jointmeters, and Tunnel Diameter Gage)	UNIAXIAL JACKING TEST (B) (Aluminum Columns, Circular Flat Jacks, REX-7P, and Tunnel Diameter Gage)	RADIAL JACKING TEST (Ring Sets, Rectangular Flat Jacks, REX-7P, Jointmeters, and Tunnel Diameter Gage)
Configuration of loaded rock area	2 - 24 in. (0.61 m) diameter shoes adjacent to each other on 3 ft (0.91 m) centers.	2 - 34.2 in. (0.87 m) diameter flat jacks, diametrically opposite each other.	Same as Uniaxial (A)	Cylinder 8 ft (2.4 m) long and 8 to 8.5 ft (2.4 to 2.6 m) in diameter. 16 each 96 x 16 in. (2.4 x 0.41 m) flat jacks, arranged around ring sets, transmit loads to rock.
Approximate total area rock surface under load	900 in^2 (0.58 m^2)	1810 in^2 (1.17 m^2)	Same as Uniaxial (A)	32,500 in^2 (21 m^2)
Maximum unit load to rock surface	800 psi (56 kg/cm^2)	1000 psi (70 kg/cm^2)	Same as Uniaxial (A)	1000 psi (70 kg/cm^2)
Maximum total load	720,000 lbs (326,800 kg)	1,810,000 lbs (821,700 kg)	Same as Uniaxial (A)	32,500,000 lbs (14,755,000 kg)
Number of drill holes instrumented	2	2	Same as Uniaxial (A)	16
Number measuring points using drill holes only	2	2	14	64
Total measuring points including tunnel diameter gage	3	3	15	66
Testing cycle	200 psi (14 kg/cm^2) - 6 days 0 psi - 1 day 400 psi (28 kg/cm^2) - 6 days 0 psi - 1 day 600 psi (42 kg/cm^2) - 6 days 0 psi - 1 day	200 psi (14 kg/cm^2) - 2 days 0 psi - 1 day 400 psi (28 kg/cm^2) - 2 days 0 psi - 1 day 600 psi (42 kg/cm^2) - 2 days 0 psi - 1 day 800 psi (56 kg/cm^2) - 2 days 0 psi - 1 day 1000 psi (70 kg/cm^2) - 2 days 0 psi - 1 day	Same as Uniaxial (A)	Same as Uniaxial (A)
Direction of applied load	Horizontal and Vertical	Any angle	Same as Uniaxial (A)	Complete 360° around test bore.
Gage length of measurements	Approximate 15 ft (4.57 m)	Approximate 20 ft (6.10 m) on each side of test adit.	14 lengths, varying from 1 to 20 ft (0.3 to 6.1 m) on each side of test adit.	64 lengths, varying from 1 to 30 ft (0.3 to 9.1 m) on each side of test adit.
Instruments measuring deformation	Joint meter Tunnel diameter gage	Same as Dual-axial	Linear variable differential transformer (LVDT) Tunnel diameter gage	Joint meter LVDT Tunnel diameter gage

FIG. 13—*Summary of jacking test equipment and procedures for measuring rock deformation moduli.*

requires a skilled geologist who is well aware of the objectives of the test program. If the test program is undertaken early in the site exploration phase, later geologic data may prove that initial test sites were not representative of the most important rock conditions. Because it is important to begin testing early, and because the knowledge of the geologic nature of the foundation evolves concurrent with the testing, it is essential that the engineers in charge of the testing program and the geologists maintain close contact. Inevitably, the scope of the program changes as new information is obtained; consequently, advance funding estimates for this work are difficult.

Once sites are chosen, the task of site preparation begins. The excavation techniques must be carefully planned to make all sites as nearly undisturbed as possible. This is necessary for the data from various test sites to be comparable and to provide an accurate determination of prototype deformations. It is also essential to the accumulation of knowledge about the behavior of widely separated rock types encountered on large engineering projects around the world. Site preparation includes the removal of rock damaged in excavations required to gain access to the site, the achievement of the proper shape and dimensions for installation of the test equipment, and the provision of a reasonably smooth rock surface for application of the test loads.

Detailed visual inspection is the principal method of determining the limits of blast damage; however, seismic and microseismic techniques may also be employed. Final preparation of test sites is done as much as possible with hand and air tools. Occasionally, rock is encountered which is simply too hard to yield to such treatment. When this happens, very light explosive charges are used in closely spaced drill holes. The explosives should have a very high velocity and a poor coupling to rapidly produce a brittle shattering of only the previously loosened material.

Following excavation of an exploratory tunnel into the rock selected for testing, the test load surface is carefully prepared. For a uniaxial test site, holes are drilled on 3 to 4-in. (7.6 to 10.1-cm) centers over an area 4 ft (1.21 m) in diameter. The depths are carefully controlled so that the bottom of each hole is in the same plane, and the rock between the holes is removed by either air tools or explosives as described above. This treatment is applied to opposing surfaces of the test drift to provide two roughly parallel surfaces approximately 10 ft (3.05 m) apart. After the removal of muck and debris, a diamond core drill is set up in the drift to drill the holes for the REX-7P's. Two NX holes are drilled 20 ft (6.1 m) deep, with their collars centered in the prepared area and on the same axis as the line of thrust of the test load. The alinement of the drill holes is done carefully so that both holes are on the same axis. This is important for later alinement with the tunnel diameter gage.

After completion of the drilling, a visual examination of the holes is done with a borehole television camera, or borescope. Major joints and fractures are located and their depths determined. This information is used in the subsequent installation of the anchors for the REX-7P. Even with careful preparations, excavation and *in situ* stresses may cause cracks near the periphery of the bore, which result in test deformations dissimilar to those produced by the prototype. Therefore, it is important that the measuring system be capable of distinguishing between closure of these peripheral cracks and closure of joints and fractures that existed before excavation of the test bore. The seven anchors are installed to span the peripheral cracks as well as the deeper fractures and joints. Thus the influence of peripheral cracks and inherent joint closure can be measured as well as the elastic deformation of intact rock between joints.

Preparations for dual-axial test sites are similar to those for uniaxial tests except that four bearing surfaces are required, although instrument holes are only provided for two.

The preparation of a radial test site usually requires enlarging a normal 5 by 7-ft (1.5 by 2.1-m) exploratory adit to a smooth-cylindrical bore 8.5 ft (2.6 m) in diameter and about 12 ft (3.6 m) long. It is accomplished by careful percussion-drilling of 1⅝-in. (4.1-cm)-diameter holes on 4 to 6-in. (10.1 to 15.2-cm) centers along the circumference and parallel to the longitudinal axis of the test bore. The holes should preferably be drilled their full length of 12 ft (3.6 m) from an enlarged drilling chamber excavated at one end of the test bore. Explosives used for this excavation employ the same criteria of high velocity and poor coupling as those used in the uniaxial test. Careful drilling and blasting of these holes will produce a relatively smooth surface having the desired configuration. Techniques for the drilling of instrument holes and the setting of anchors for the REX-7P are similar to procedures described above for the uniaxial tests, except that many more holes and anchors are required.

Data and Conclusions

The reduction of data to derive the linear deformation from the readings taken with both the special joint meter and the REX-7P requires only the application of a calibration factor. The calibration factor is determined by measuring linear length changes over the range of the instruments, utilizing a precise micrometer, and plotting these versus the output. The tunnel diameter gage measures the change in tunnel dimensions directly. It is, of course, prudent to occasionally check its dial indicator readings against a standard known length. Determination of the deformation modulus from the dual-axial jacking test equipment and the uniaxial tests was accomplished by utilizing the procedure which follows.

An equation for the normal deflections at any point in the interior of a semi-infinite elastic solid bounded by a plane surface[3] is given by

$$W = \frac{P}{2\pi E} \left[(1 + \mu) Z^2 (r^2 + Z^2)^{-3/2} + 2(1 - \mu^2)(r^2 + Z^2)^{-1/2} \right] \quad \text{...(1)}$$

where:

r = radius measured parallel to the loaded surface,
Z = depth below the surface of the solid,
P = concentrated load applied perpendicular to the loaded surface,
E = modulus of elasticity,
μ = Poisson's ratio, and
W = deflection of point having radius r from line of load and depth Z from plane of test surface.

1. *Dual-axial Test*—In the case of the dual-axial test, the condition of a uniform vertical pressure is used. The term E is replaced with E_d because, as noted in the introduction, jointed or fractured rock masses are not truly elastic. These conditions were applied to Eq 1 by Boussinesq to obtain Eq 2 below, which may be readily solved by utilizing the deflection at the center of each pad as measured by the embedded joint meters:

$$E_d = \frac{q}{W} (25.7 - 0.78\mu - 26.5\mu^2) \quad \text{.........................(2)}$$

where:

W = deflection at center of pad for gage length of 15 ft,
E_d = modulus of deformation, and
q = intensity of uniform distributed load.

The tunnel diameter gage for the dual-axial test is located midway between the two jacks, or 1.5 ft (0.45 m) from the center of each loaded area. In order to solve for the modulus, two assumptions must be made: (1) the rock at the top and bottom of the jacks behaves identically and (2) the rock movement caused by each jack is equal at the point midway between them. Both of these assumptions are subject to question; and since no embedded instrument is installed in the arch, there is no way to verify either. The equation for the modulus when the displacement is measured by the tunnel diameter gage is in the form of an elliptical integral which may be reduced to

$$E_d = 34.2(1 - \mu^2) \frac{q}{W} \quad \text{................................(3)}$$

where W is now the total rock deformation as taken from tunnel diameter gage.

2. *Uniaxial Tests*—The uniaxial tests used the circular flatjacks with a

hole in their centers. If we call the outside diameter of the annulus a_2 and the inside diameter a_1, the equation for E_d becomes

$$E_d = \frac{q}{W}\left(30.23 - 30.82\mu^2 - 0.60\mu \right) \ldots\ldots\ldots\ldots (4)$$

Where W, the deflection at the center of the pad, is determined from the REX-7P having a gage length of 20 ft,

$$E_d = 32(1-\mu^2)\ \frac{q}{W} \ldots\ldots\ldots\ldots (5)$$

The solution for the modulus from the tunnel diameter gage readings in the uniaxial test utilizes Eq 5, since the points of measurement are the same as with the embedded instrumentation. However, it is necessary to assume that the rock on both ends of the test setup behaves identically, that is, W equals the tunnel diameter gage reading divided by two. This assumption is frequently invalid; however, the tunnel diameter gage provides a true check on the REX-7P as its movement should equal the sum of the opposing REX-7P's plus the small correction for deformations beyond the bottom anchors.

3. *Radial Jacking Test*—The analysis of data from the radial jacking tests involves the use of a mathematical model corresponding to the configuration of the field test. The mathematical model used was an infinitely long, cylindrical hole in an infinite homogeneous mass. This solution and its application to the radial test are described by Dewayne Misterek in the paper "Analysis of Data from Radial Jacking Tests," which can be found on pp. 27–38 in this volume.

As previously mentioned, the calculation of moduli is based solely on elasticity theory. It must be realized that this represents, at best, an approximation since the rock deformations include inelastic, as well as elastic, movements that are the result of the characteristics of the material and its discontinuities. Furthermore, the induced peripheral cracks may cause the distribution of stress with depth to differ significantly from that assumed by elasticity theory, and for this reason, research is being initiated to measure stress, as well as deformation, beyond the loaded surface. To assure a continuing awareness of these inelastic factors, the term "modulus of deformation" (E_d) has been used when the elastic equation is used to reduce *in situ* data. A typical representation for the total deformation which utilizes the dual-axial, uniaxial, and radial jacking tests is shown in Fig. 14. This deformation-load-time curve shows deformation, rebound, and set under various loading conditions.

Utilizing the REX-7P's a depth-deformation curve shown in Fig. 15, is obtained. This curve gives the change in deformation at various depths

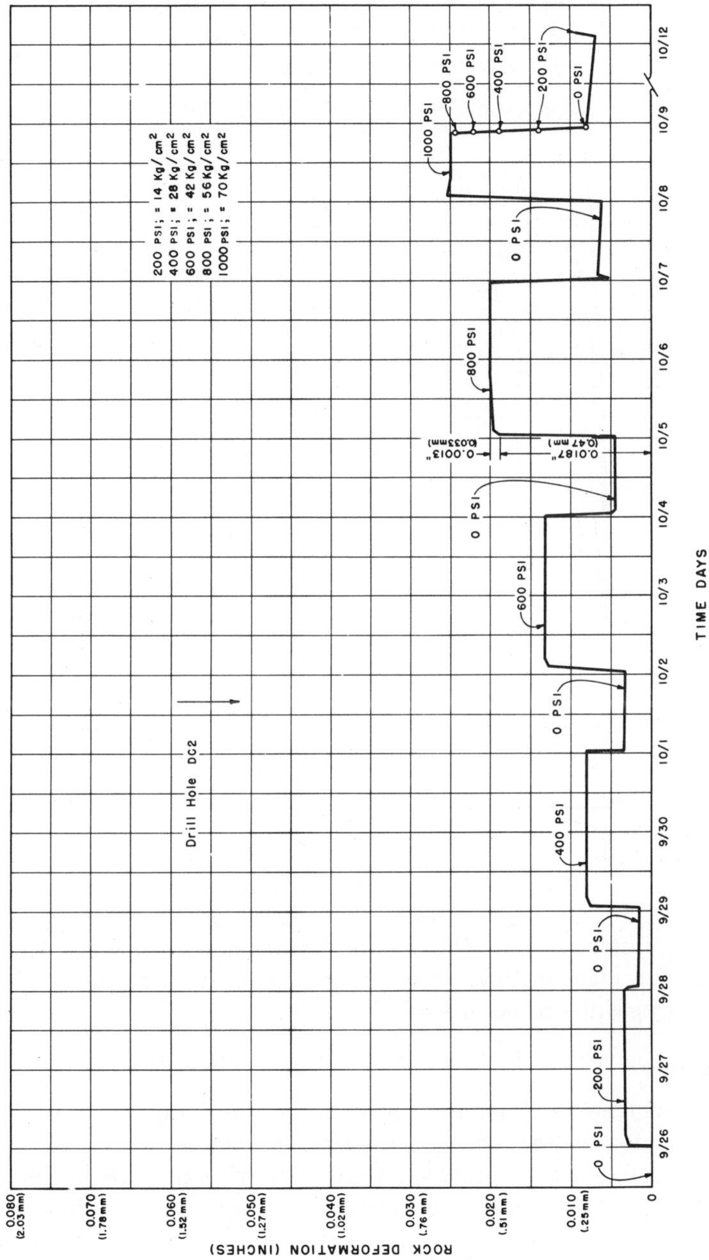

FIG. 14—*Rock deformation versus time.*

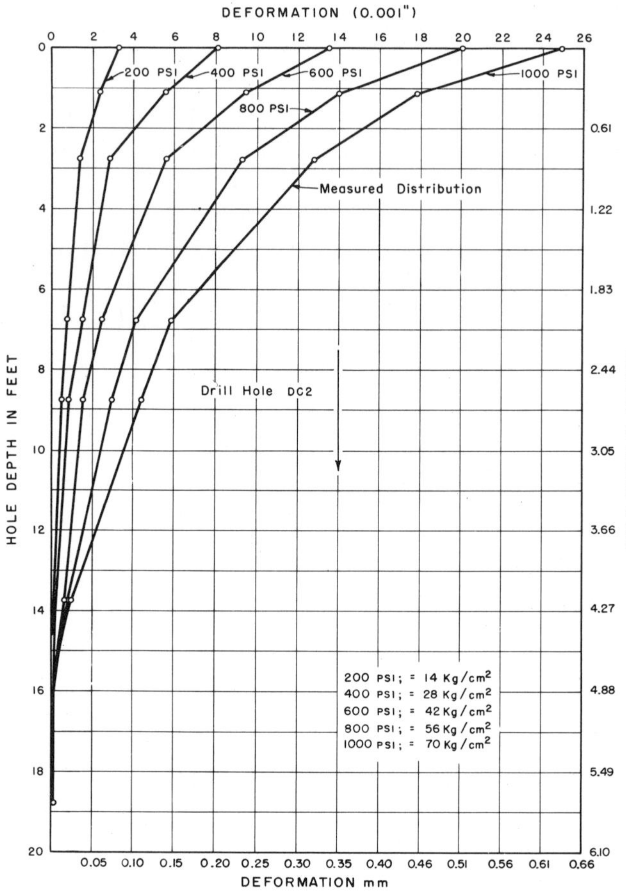

FIG. 15—*Hole depth versus rock deformation.*

along the instrumented hole. Figure 16 illustrates the total deformation measured in eight instrument holes around the central plane of a radial jacking test with various loads.

Summary

Improvements in techniques, procedures, and hardware have facilitated large-scale foundation testing programs in remote areas. To achieve a practical and realistic estimate of the deformation modulus for a large, heavily loaded foundation, several different test methods have been developed. Numerous cores of intact rock from holes blanketing the area are tested in the laboratory, and the data are combined with results of several axial and radial jacking tests which show the influence of discontinuities on prototype deformation.

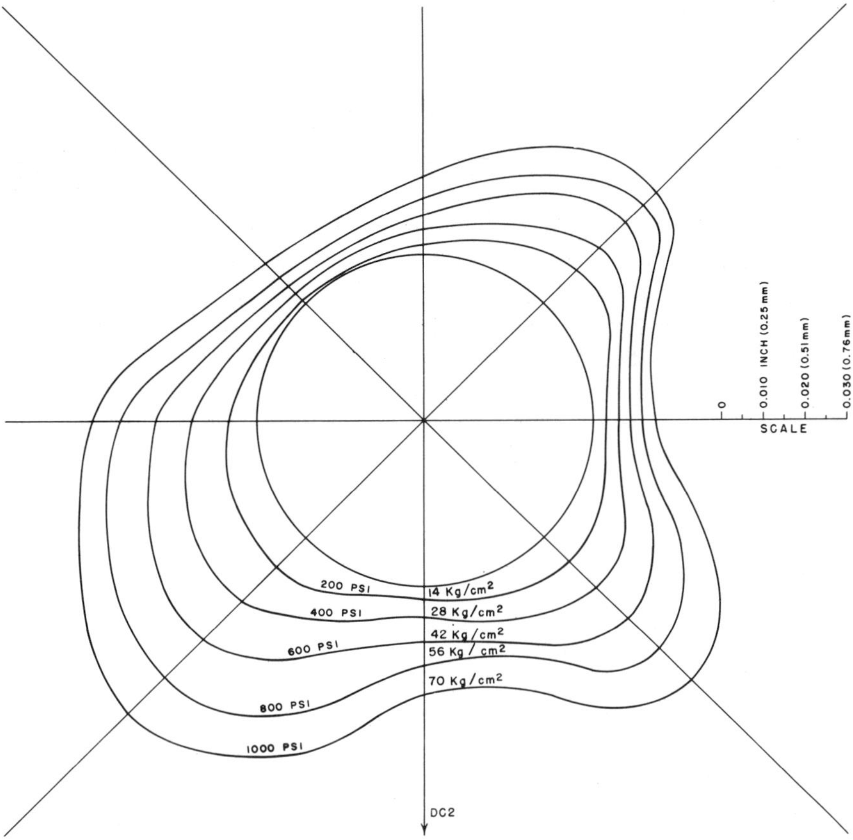

FIG. 16—*Deformation around the center section of the radial jacking test.*

Dual-axial tests are not presently being used because of their weight and the inability to make tunnel diameter measurements between the centers of the loaded areas. However, for the loading of highly deformable rock, the hydraulic rams used in this method might be more practical than flatjacks.

The lightweight uniaxial test may be used in numerous locations throughout the foundation, because it is inexpensive in comparison to the larger radial test and requires only a few weeks to prepare and run. It may be used at points considered representative of large portions of the foundation or at a number of sites containing special rock types. The radial jacking test equipment required an investment of about $50,000, and hence its application would usually be reserved for those rock types which compose the biggest part of the foundation or at sites which are apt to critically influence the prototype deformation.

The most significant advancement to date toward the improvement of techniques for determining the deformation modulus *in situ* has been the development of instrumentation that is capable of continuously sensing and recording numerous deformations within a loaded rock mass and that is able to distinguish the following factors:

1. Closure of cracks near the periphery of the test bore, including cracks caused by blasting, relief of *in situ* stresses, and possibly radial cracks induced by the test load configuration.
2. Closure of fractures and joints that existed before construction of the test bore, which often contributes much more to prototype deformation than Item 3 below.
3. The elastic deformation of intact rock.

The most significant technique remaining to be developed is that required to measure the distribution of stress with depth. Regardless of the care taken in excavating and preparing a test site, some induced peripheral cracking will occur which will affect the distribution of stress to some extent. The REX-7P's are capable of determining the extent of the influence of these cracks on the deformation at depth. If the stress can be measured beyond that point, a more realistic estimate of the prototype deformation can be achieved.

References

[*1*] Soil Mechanics and Foundation Engineers, "Report on *In Situ* Tests, Dworshak Damsite," Shannon and Wilson, Inc., Seattle, Wash., Dec. 1964.

[*2*] Wallace, G. B., Slebir, E. J. and Anderson, F. A., "Radial Jacking Test for Arch Dams," U.S. Bureau of Reclamation, 10th Rock Mechanics Symposium, Austin, Tex., 20–22 May 1968.

[*3*] Timoshenko, S. and Goodier, J. N., *Theory of Elasticity,* McGraw-Hill Book Co., Inc., New York, 1951.

D. L. Misterek[1]

Analysis of Data from Radial Jacking Tests

REFERENCE: Misterek, D. L., "**Analysis of Data from Radial Jacking Tests,**" *Determination of the* In Situ *Modulus of Deformation of Rock, ASTM STP 477,* American Society for Testing and Materials, 1970, pp. 27–38.

ABSTRACT: The radial jacking test consists of applying a radial pressure to the surface of a short length of tunnel and then measuring the resulting displacements that occur within the rock mass. These measured values are used to determine a load-displacement relationship suitable for the design of important structures. The method presented for obtaining this relationship is based on solving for stresses and displacements in an infinite elastic body which contains a circular hole loaded radially along a finite portion of its length. The solution is used to obtain data for a curve which can then be used to determine the deformation moduli indicated by a radial jacking test. Studies conducted indicate that small errors in determining Poisson's ratio do not seriously affect moduli calculations and that the optimum ratio of loaded length to diameter of opening is approximately unity. Curves are also presented which indicate how moduli calculations should be modified if tangential tensile stresses resulting from radial loads produce radial cracking in the material immediately around the tunnel.

KEY WORDS: rock mechanics, foundation investigation, dam foundations, design data, deformation modulus, data analysis

In situ deformation tests are performed to obtain information necessary for the rational design of arch dams, pressure tunnels, or other structures whose design is dependent upon the deformation characteristics of a rock mass. This necessary information consists of numerical values and expressions which can be used to predict the behavior of the rock mass when loads from the structure are applied. Therefore, the problem becomes one of applying known loads, measuring the resulting displacement, and then analytically relating these loads and displacements as a means of predicting the displacements that will be produced by structural loads.

[1] Research structural engineer, Division of Design, Bureau of Reclamation, Denver, Colo. 80225. Personal member ASTM.

Throughout the world numerous techniques are employed for performing *in situ* deformation tests. These include uniaxial jacking tests, pressure chamber tests, and radial jacking tests, all of which are performed in tunnels. (Another technique, in which the sides of a precut slot are loaded with flatjacks, can be used on any surface.) This discussion will be concerned only with the radial jacking deformation test. Test equipment and procedures used to perform the test are described elsewhere [*1*][2] and will be mentioned only briefly here. This paper discusses an analytical procedure for converting test data into deformation moduli. The analytical procedure is then used to investigate the effect of some of the variables that are inherent in this type of test.

The radial jacking test performed by the U. S. Bureau of Reclamation consists of the application of uniform pressures up to 1000 psi (70.3 kg/cm^2) to the surface of a portion of tunnel 8 ft (2.44 m) in length and approximately 8 ft (2.44 m) in diameter. Relative displacements between the rock surface and points at various depths within the rock are measured in eight equally spaced radial holes in the central longitudinal section and in four equally spaced radial holes at each of the quarter points of the loaded length. As an additional check, changes in tunnel diameter are measured between reference points in the central section.

The radial jacking loading system was first used by Tiroler Wasserkraftwerke, A. G., (TIWAG) Innsbruck, Austria, to gather data to aid in the design of pressure tunnels. For this application changes in tunnel radius were the only displacement values measured. The measured displacements were converted to deformation moduli by using the solution from the theory of elasticity for an infinitely thick cylinder subjected to internal pressure; but it was recognized that the test arrangement loaded only a short length of the tunnel and, therefore, did not completely satisfy the conditions of the analytical solution. Lauffer and Seeber [*2*] describe the technique used by TIWAG to circumvent this difficulty: by measuring radial displacements both inside and outside the loaded length and then applying Maxwell's law of reciprocal deflections, it is possible to estimate the surface deflections that would have occurred if the loaded length had been infinitely long.

The radial jacking test employed by the Bureau of Reclamation has several important differences from the Austrian tests. In the first place, data from Bureau tests are being obtained primarily for use in the design of concrete arch dams. Secondly, the displacement measurements made at depth within the rock mass provide additional information that can be used to evaluate the modulus of the rock at depth or the effect of jointing. The condition of the rock, the distribution and direction of applied load, and

[2] Italic numbers in brackets refer to the list of references at the end of the paper.

the stresses within the rock mass during the radial jacking test are similar to the corresponding quantities for a loaded pressure tunnel; therefore, measured radius changes provide a good basis for determining the proper load-displacement relationship for the surface of the tunnel. However, if we are to fully utilize the additional information provided by the more sophisticated system for gathering displacement data, and if the calculated moduli are to be the best possible values for use in the design of a concrete arch dam, a slightly different approach is required.

The determination of deformation moduli for dam design requires special attention, because the stress pattern (and, therefore, the rock behavior resulting from the radial test) is different from that produced by the structure. During the radial jacking test, application of a radial pressure on the walls of a circular opening produces not only a radial compressive stress but also a tangential stress which is tensile in nature. Depending upon the initial *in situ* stress conditions and the manner in which the natural stresses are redistributed as a result of the excavation of the tunnel, the rock in the tunnel walls may be subjected to tangential tensile stresses which can produce radial cracking. A stress field of this type and the associated rock behavior are not completely representative of conditions produced by loads from an arch dam. For a dam, a zone of rock that has been disturbed by blasting and probably has a reduced deformation modulus exists near the abutment and foundation surfaces. However, the relative effect that this zone has on the displacements produced by a dam is much less than the relative effect that blast damage and possible radial cracking have on deflections produced by radial jacking tests. Therefore, it is essential that we have a procedure for evaluating test data from a loosened, and possibly cracked, zone or from the material outside the disturbed zone. The resulting information can then be used to evaluate the probable action of the rock mass when subjected to the structural loads. Using deflection measurements that are obtained from radial jacking tests performed by the Bureau, the following technique can be applied for the evaluation of deformation moduli that are applicable to the design of either pressure tunnels or arch dams.

The basis for this technique is an analysis of the stresses and deflections around a circular hole in an infinite elastic medium, which are produced by a uniform pressure applied to a finite length of the hole. An exact solution to this problem has been presented by Tranter [*3*], and all results and curves presented here are based on that solution. As this solution was developed using the theory of elasticity, the basic assumptions of continuity, homogenity, isotropy, and elasticity have been made although it is recognized that no rock mass completely satisfies these requirements. However, if joints, cracks, and other planes of weakness do not open or fail by shear, the action of the rock mass may approach that of an elastic body for moderate load intensities. The assumption of elasticity is also made less critical by the manner in which the data are used. When the test is performed, load and displacement are measured, and a value is calculated to express the relationship between the two variables. For design, the analytical process is reversed, and displacements are calculated from known loads and the previously determined load-displacement relationship.

Tranter's solution expresses the radial displacement and the components of the stress field in terms of improper integrals of the first kind (upper limit is infinite) of various combinations of the Bessel functions K_0 and K_1, common trigonometric functions, physical properties of the material, load intensity, and dimensions and loaded length of the hole. These complex expressions could not be integrated in closed form; therefore, a computer program was prepared to numerically evaluate the integrals.

As in any approximate solution, care must be taken to insure that the final result is actually a good approximation of the true value. In the approximate numerical evaluation of these integrals, three sources of possible error exist:

1. Size of the interval between the ordinates used in the numerical integration may influence the size of the error.

2. Truncation of the integration at a finite upper limit may introduce an error by neglecting the portion of the integral beyond that upper limit.

3. The Bessel functions were evaluated from polynomial approximations [*4*]. Inherent inaccuracies in these relations may be magnified in making the required combinations of the functions.

The following steps were taken to insure that the magnitude of error was permissible:

1. Two computations were carried out which were identical except that the interval between ordinates in one case was one half of that in the other. Agreement between the two sets of results indicates that the interval was sufficiently small. A smaller interval was used for the small abscissas where values of the Bessel functions change more rapidly.

2. The effect of truncating the integrals at a finite upper limit was evaluated by successively extending the upper limit of integration until any additional contribution to the numerical value of the integrand was insignificant. A check on the general method, and the value of radial stress in particular, was obtained from the numerical evaluation of the radial stress at the surface of the opening and within the loaded area. This value, which should be equal to the applied load, agreed to four significant figures.

3. The polynomial approximations for Bessel functions of large arguments involve a multiplication by the negative exponential. During subsequent combinations of the various Bessel functions, a considerable loss of significant figures was experienced. This difficulty was satisfactorily avoided by

postponing the multiplication by the exponential until the last possible opportunity.

By utilizing this procedure, the radial, tangential, longitudinal, and shear stress and the radial displacement at any point within an elastic mass subjected to the specified boundary condition can be evaluated.

Results

Although similar data can be determined for any location, results will be presented only for points in the plane normal to the longitudinal center line of the opening at the midpoint of the loaded length. The data for all curves were computed for an opening with a diameter of 8 ft (2.44 m). All constants used for the calculations are noted on the figures.

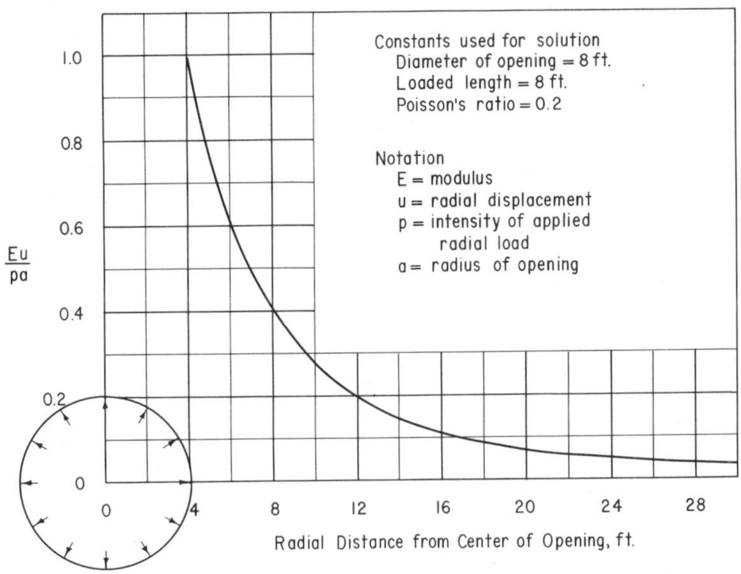

FIG. 1—*Radial displacement produced by radial jacking test.*

Typical results for a given set of conditions are presented in Figs. 1 and 2. These curves show dimensionless parameters which depict the variation in radial displacement or stress intensity at different distances from the surface of the opening. Data of the type illustrated in Fig. 1 are used to determine deformation moduli from displacement measurements made during radial jacking tests. The ordinates of the curve are the values of the dimensionless quantity, Eu/pa, where E is the modulus, u is the radial displacement, p is the intensity of applied radial load, and a is the radius of opening. Indicated deformation moduli can be easily calculated from the

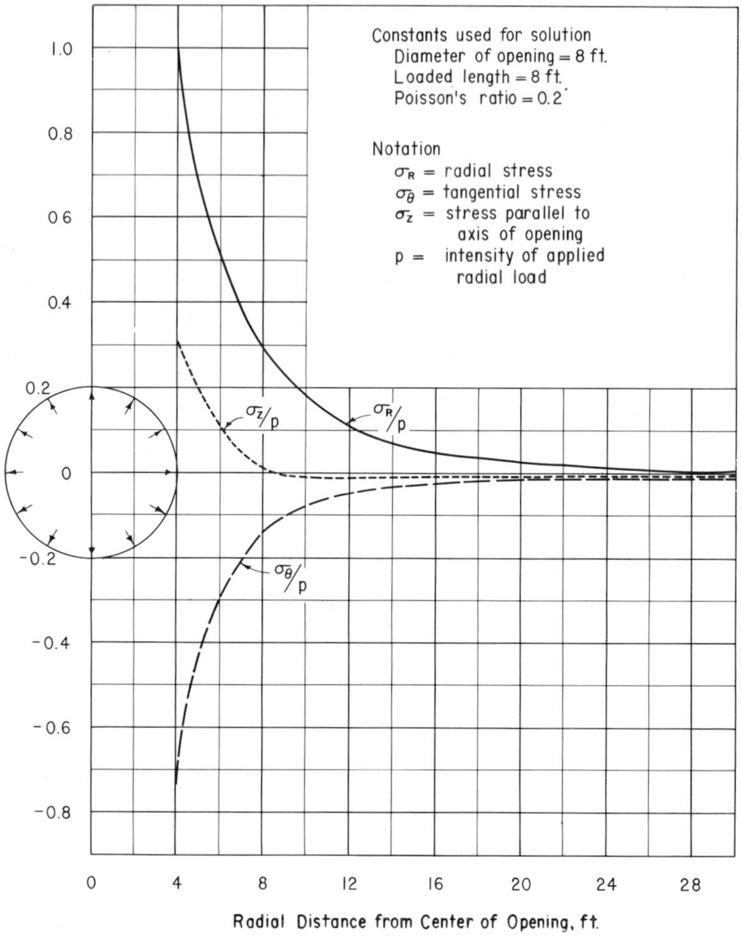

FIG. 2—*State of stress produced by radial jacking test.*

known test constants, the measured difference in displacement between two points, and the difference in the Eu/pa values between the same two points.

For example, assume that the relative displacement between the surface of the opening and a point 12 ft (3.66 m) in depth is measured to be 0.0094 in. (0.24 mm) when a radial pressure of 600 psi (42.18 kg/cm^2) is applied to the 8-ft-diameter (2.44 m) opening. The value of Eu/pa at the surface of the opening, as read from the curve in Fig. 1, is 0.996, while at 12 ft (3.66 m) from the surface, Eu/pa is 0.115. Therefore, the radial displacement at the surface, u_0, is 0.996 pa/E, and u_{12} is 0.115 pa/E. The relative displacement between two points, $u_0 - u_{12}$, is 0.881 pa/E. By substituting the measured relative displacement, 0.0094 in. (0.24 mm), and the known test constants, the indicated deformation modulus, E, becomes $0.881 \times 600 \times 48 \div 0.0094$, or 2,700,000 psi (190,000 kg/cm^2). The curve shown in Fig. 2 is not used for calculating moduli but does illustrate the manner in which stresses decay with depth for these loading conditions.

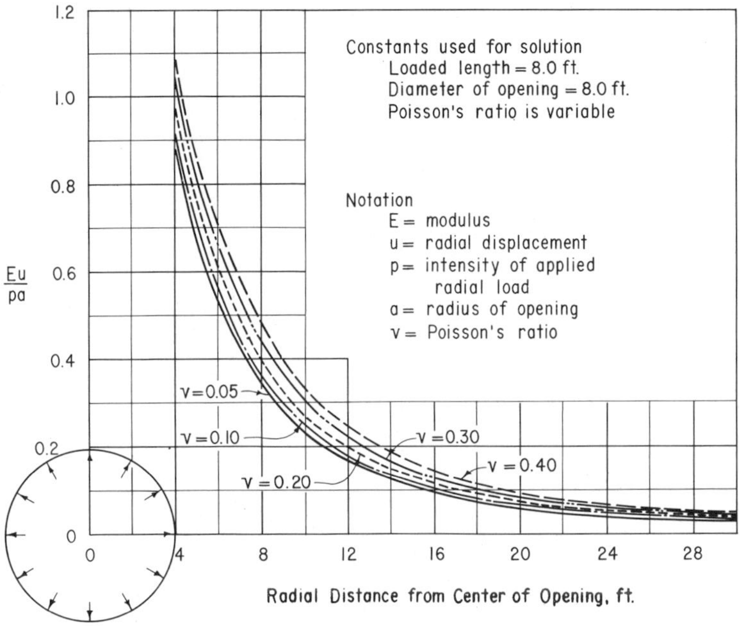

FIG. 3—*Effect of Poisson's ratio on radial displacement.*

Figures 3, 4, 5, and 6 show results from typical studies conducted to determine the effect of some of the variables that cannot be precisely evaluated or those that can be controlled.

Currently, the value of Poisson's ratio is determined from laboratory tests on intact rock core specimens, but there is no assurance that the effective Poisson's ratio of a jointed rock mass is equal to the value indicated by tests on cores. The curves shown in Fig. 3 illustrate how calculated radial displacements are affected by changes in Poisson's ratio. These curves indicate that the maximum error in the calculated modulus will be approximately 10 percent if the value of Poisson's ratio is determined to an accuracy of $\pm$ 0.1. Normally, Poisson's ratio will be known to an accuracy better than $\pm$ 0.1; therefore, errors resulting from inaccuracies in its determination will be of acceptable magnitude.

Figures 4 and 5 show how variations in loaded length affect the amount of displacement that occurs. Data of this type are important for the planning and design of radial jacking test equipment. If the loaded length is

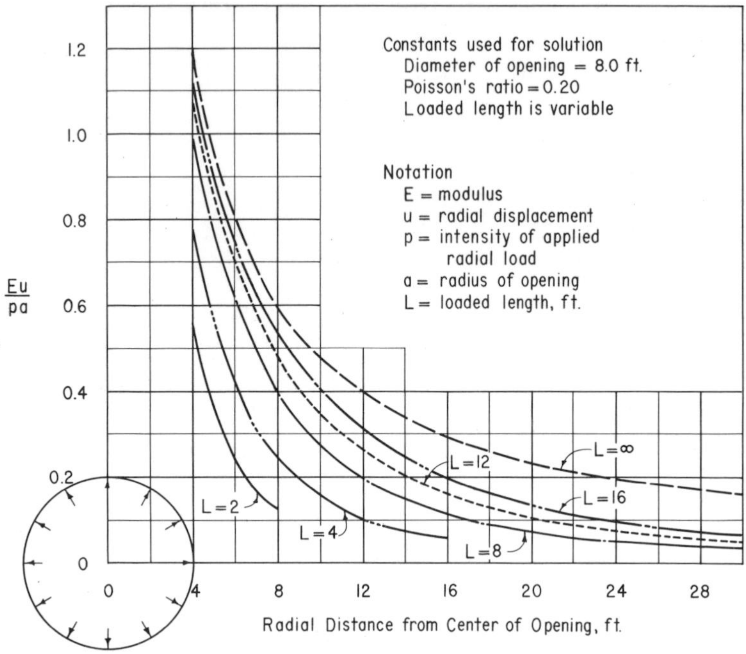

FIG. 4—*Effect of loaded length on radial displacement.*

too short, the volume of rock subjected to stress might be too small to accurately reflect the properties of the rock mass. Also, the amount of displacement would be much less, and inherent inaccuracies in the measuring system might become signficant in relation to the quantity being measured. On the other hand, an excessively long, loaded length would be uneconomical because of the requirement for additional loading equipment and site preparation. Figure 5 shows that the magnitude of the surface deflection increases rapidly as the loaded length is increased to a value about equal to the diameter of the opening but that additional increases in loaded length produce only small increases in radial displacement.

As mentioned previously, a zone of disturbed rock surrounds every tunnel, the zone being especially pronounced for those tunnels driven by blasting methods. Adding to this inhomogeneity caused by construction, the forces applied to the rock by the radial jacking test produce tangential tensile stresses which can lead to radial cracking around the opening. These two causes produce a zone of material around the opening whose behavior can deviate greatly from that of a continuous, homogeneous, elastic material. The curves in Fig. 6 give an indication of how seriously this deviation would affect the calculations of deformation modulus. Values for these curves were calculated using the assumption that the rock mass

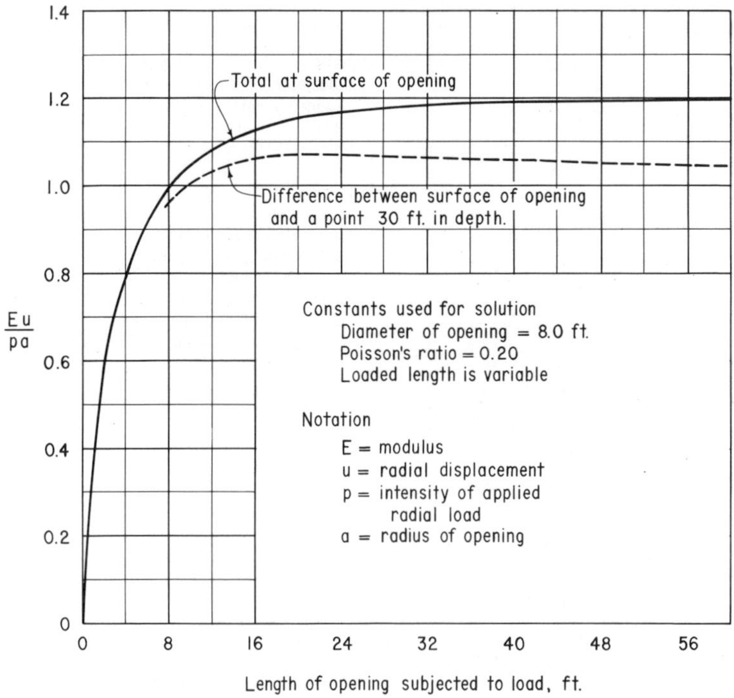

FIG. 5—*Effect of loaded length on radial displacement at surface of opening.*

contained radial cracks moving outward for a known distance from the surface of the opening. The material containing the radial cracks was then assumed to act as a column of rock loaded in uniaxial compression and free to expand laterally. Strain in the rock column was computed for a uniform stress distribution across the area of the column. The material outside the radially cracked zone was considered to be a continuous, homogeneous, elastic mass. The total force applied to the inside of the opening was assumed to be transferred through the rock columns and applied to the larger diameter surface; thus, the load intensity was reduced by a corresponding amount because the total force was now distributed over a greater area. Using these assumptions, the values for the curves in Fig. 6 were computed for radial cracking extending outward 0.125, 0.25, 0.375, 0.5, 0.75, 1, and 1.5 diameters from the surface of an 8-ft-diameter (2.44 m) hole.

An analysis of the data in these plots reveals the magnitude of the differences that might be introduced into calculated moduli when the material is erroneously assumed to be uncracked. As would be logically expected, the discrepancies increase as the cracked zone penetrates deeper into the rock mass. When the depth of the cracked zone is one fourth of

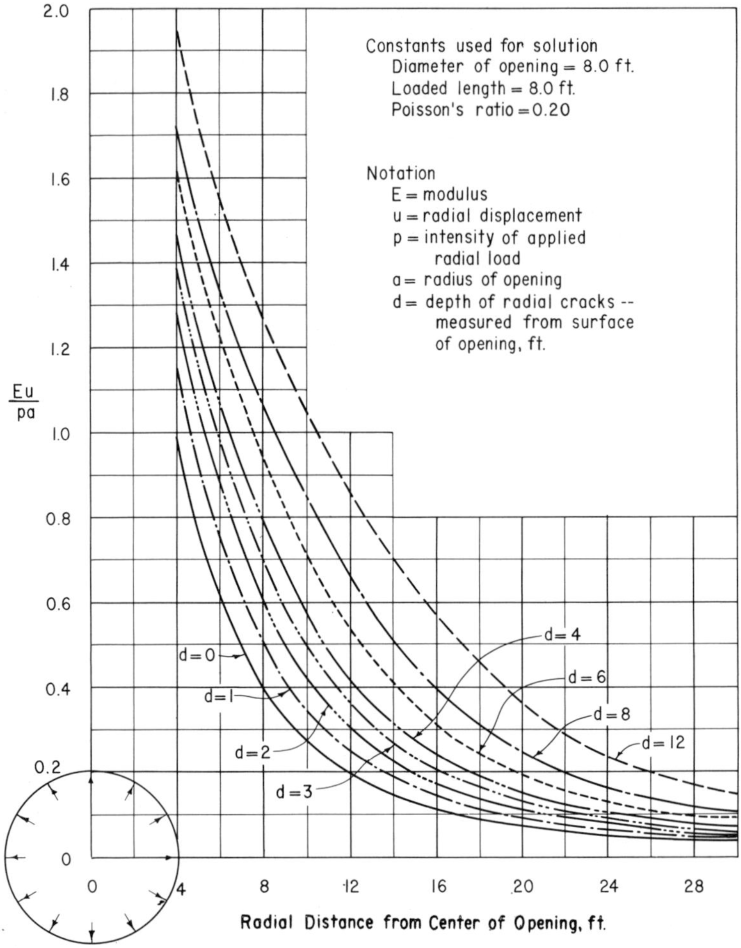

FIG. 6—*Effect of radial cracking on radial displacement.*

the radius of the original opening (1 ft or 0.30 m in this example), the maximum difference would be about 25 percent. The maximum difference increases to nearly 50 percent when the crack depth is one half of the original radius and up to 100 percent for a crack depth equal to the initial radius.

The depth to which radial cracking will penetrate depends upon four factors: (1) the amount and direction of natural joints and cracks prior to loading, (2) the tensile strength of the rock mass in a tangential direction, (3) the intensity of the applied load, and (4) the *in situ* stress field, which determines the initial stress conditions in the rock prior to loading and the manner in which the *in situ* stress distribution is altered by construction activities.

Results obtained to date indicate that frequently the zone of material immediately around the opening reacts to load differently than the material at greater depth. Figure 7 shows typical results of modulus calculations made for deformations measured during the initial test operation of the radial jacking equipment. The data indicate that a zone of disturbed material extends to a depth somewhere between 1.1 and 2.5 ft (0.34 and

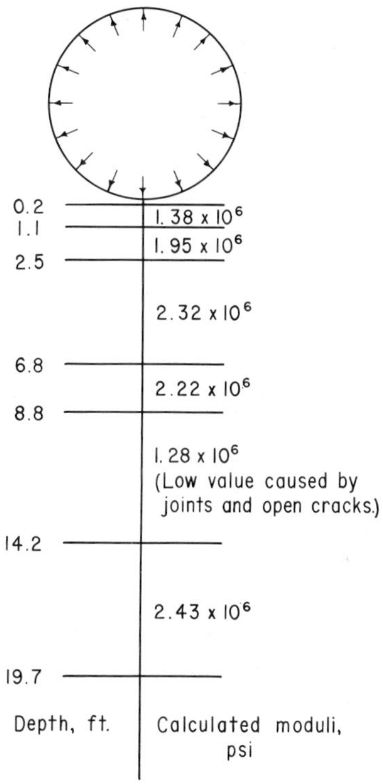

FIG. 7—*Variation in calculated modulus with distance from opening.*

0.76 m) from the surface of the opening. It might be argued that this zone of reduced modulus is the result of blast damage and not radial cracking; however, even if this is true initially, blast-damaged rock would almost certainly not possess the tensile strength needed to withstand the applied loads and would therefore crack to that depth. By using the original data to determine the depth at which an apparent discontinuity in the modulus exists, the depth of the disturbed or radially cracked zone can be estimated. Then, the modulus calculation can be repeated utilizing the most appropriate curve from Fig. 6. This procedure should provide more realistic values of the deformation modulus for use in the design of arch dams or other structures that apply compressive loads to the rock mass.

Conclusions

1. A computer program has been prepared for the numerical evaluation of the stresses and displacements indicated by the exact solution from the theory of elasticity for a uniform pressure applied to a finite length of an infinitely long circular opening in an infinite elastic medium. Numerical results of this solution provide the basis for the evaluation of deformation moduli from radial jacking tests.
2. Studies indicate that, even though Poisson's ratio for a rock mass cannot be precisely evaluated, resulting errors in calculated moduli will not be excessive.
3. The optimum length of a tunnel subjected to load is approximately equal to the tunnel diameter. This loaded length provides displacements of 80 to 85 percent of the maximum value that would be obtained for an infinitely long, loaded length and still keeps site preparation and test equipment requirements within reasonable limits.
4. The zone of disturbed or radially cracked rock that normally surrounds a loaded test site has a significant effect on displacements. Consequently, moduli calculated from measured displacements may include important errors unless the proper adjustments are made. A method is proposed for determining more realistic values of deformation moduli under these conditions.

Acknowledgments

The procedure for the numerical evaluation of the mathematical expressions for stress and displacement was developed by W. T. Moody, U. S. Bureau of Reclamation.

References

[*1*] Wallace, G. B., Slebir, E. J. and Anderson, F. A., "Radial Jacking Test for Arch Dams," 10th Symposium on Rock Mechanics, May 1968, in press.
[*2*] Lauffer, H. and Seeber, G., "Design and Control of Linings of Pressure Tunnels and Shafts, Based on Measurements of the Deformability of the Rock," Question 25, Report 91, Seventh Congress of Large Dams, Rome Italy, 1961.
[*3*] Tranter, C. J., *Quarterly of Applied Mathematics,* QAMAA, Vol. 4, No. 3, 1946, pp. 298–302.
[*4*] *Handbook of Mathematical Functions,* National Bureau of Standards, Applied Mathematics Series 55, NBAMA, Section 9.8, June 1964, pp. 378–379.

M. Rocha[1]

New Techniques in Deformability Testing of *In Situ* Rock Masses

REFERENCE: Rocha, M., "**New Techniques in Deformability Testing of *In Situ* Rock Masses,**" *Determination of the* In Situ *Modulus of Deformation of Rock, ASTM STP 477,* American Society for Testing and Materials, 1970, pp. 39–57.

ABSTRACT: Two new techniques developed in recent years by the LNEC for the *in situ* determination of deformability in rock masses are presented. One of them allows the determination of deformability along boreholes to depths exceeding 200 m by means of a dilatometer which was developed to operate in holes 7.6 cm in diameter that may contain water. The other technique allows the testing of large volumes of rock mass by means of thin flatjacks installed in slots cut in the rock mass by a diamond disk. Results of the application of these new techniques showing their reliability and accuracy and emphasizing their practical advantages are presented.

KEY WORDS: modulus of elasticity, rocks, rock properties, tests, deformation, boreholes, foundations, evaluation

Although rock masses generally have a small deformability, the behavior of rigid structures, such as concrete dams and tunnel linings, may depend greatly on the displacements undergone by the rock masses.

As is well known, it is impossible in most cases to characterize the deformability of rock masses using laboratory tests, because to be representative of the fracturing and heterogeneity usually occurring in rock masses, the specimens would have to be excessively large. Even the usual *in situ* jackloading tests do not yield satisfactory results in most cases, because the rock mass in the tested zone is disturbed by the excavation, and also because the tested volumes are not yet representative of the rock mass.

To avoid these difficulties a new, thin flatjack technique (described later) was developed, which allows the quick testing of large undisturbed vol-

[1] Director, Laboratorio Nacional de Engenharia Civil (LNEC), Lisboa, Portugal.

umes.[2] Furthermore, the determination of the deformability of rock masses along boreholes, even during the preliminary geological studies, is recognized to be of great interest. Thus, a dilatometer has been built to operate in holes having a diameter of about 76 mm, as usual in site investigations.[3]

In order to assess the deformability of a rock mass it is considered advisable to (1) determine the deformability with a dilatometer, (2) establish a zonation of the rock mass according to the results obtained, (3) determine the deformability of each zone by means of tests on large volumes with the new, thin flatjack technique, and (4) establish a correlation between the dilatometer test results and those of the flatjacks in an attempt to reduce the number of the latter and also extend the information supplied by them to the zones of the rock mass only tested with the dilatometer.

Determination of Deformability by Means of Dilatometer

The dilatometer that was developed was designed to have the ability to (1) be used in boreholes of the usual diameter in site investigations, (2) operate in holes that may have water, (3) be used down to great depths of even more than 200 m, (4) apply radial pressures up to 150 kg/cm^2 on the borehole wall, and (5) enable the measurement of deformations of four diameters at 45 deg with one another to obtain information about the anisotropy of the rock mass. Besides, it can also be used in holes drilled in any direction, even the horizontal.

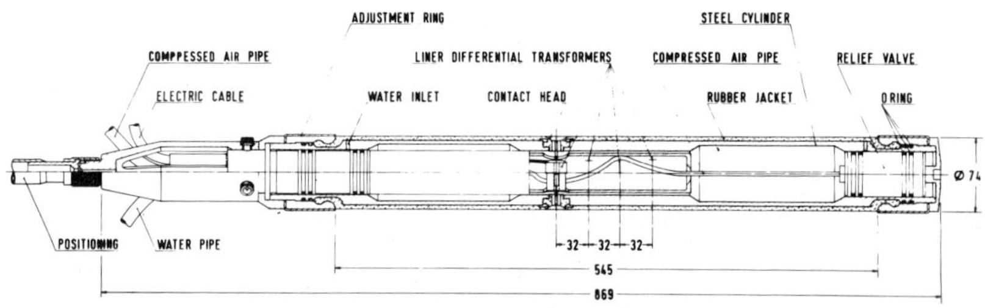

FIG. 1—*Longitudinal section of dilatometer.*

Figure 1 shows schematically a longitudinal section of the dilatometer as it is constructed now. The external diameter of the apparatus in the zone that applies the pressure is 74 mm, which corresponds to a clearance of

[2] Rocha, M., "Rock Mechanics in Portugal," *Proceedings of the First Congress of the International Society of Rock Mechanics,* 25 Sept.–1 Oct. 1966, Lisbon, Portugal, 1967, Vol. 3, pp. 121–132.

[3] Rocha, M., Silveira, A., Grossman, N. and Oliveira, E., "Determination of the Deformability of Rock Masses Along Boreholes," *Proceedings of the First Congress of the International Society of Rock Mechanics,* 25 Sept.–1 Oct. 1966, Lisbon, Portugal, Vol. 1, 1967, pp. 697–704.

2 mm in relation to the diameter of the borehole; the diameter of the ends can be changed by means of rings with diameters adequate to the actual diameter of the borehole. By pumping water between the rubber membrane and the steel cylinder, the wall of the hole is submitted to a uniform radial pressure over a length of 54.5 cm. Once a test has been performed, to take the dilatometer out or to place it in a new test position one must first take out the water that had been introduced to apply the pressure. This is done by a relief valve operated by compressed air. The dilatometer is introduced into the borehole connected to steel rods which allow the depth and azimuth of the apparatus to be known.

To measure the deformations undergone by the four diameters at 45 deg, differential transducers with a measuring range of about 5 mm and an accuracy of about 1 μm are used. The coil and core of each transducer are connected to steel rods, the ends of which establish the contact with the borehole wall.

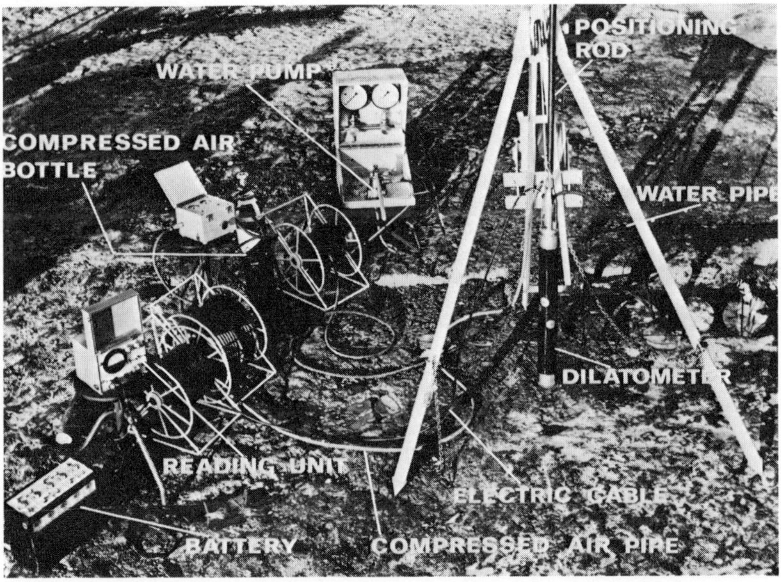

FIG. 2—*Test equipment in position around borehole.*

Figure 2 shows the dilatometer, which weighs 12 kg, being lowered into a borehole and the equipment needed for its operation. When the borehole wall is very rough or likely to disintegrate, the hole is grouted without pressure and drilled again. In order to limit the penetration of grout into the rock mass and therefore reduce the disturbance of the deformability, the grouting should be carried out in short sections about 10 m in length of the borehole. When it is desired to determine the effect of a treatment, this should be applied along the borehole, simulating the conditions that will prevail in practice.

The accuracy and reliability of the measurements made with the dilatometers manufactured by the Laboratorio Nacional de Engenharia Civil (LNEC) have been assessed by tests carried out inside steel pipes. In these tests deformations read by the dilatometers are compared with the corresponding external diameter deformations, also read by differential transducers (Fig. 3). Very good agreement has been obtained. The program of the tests to be performed with the dilatometer depends on the problem to

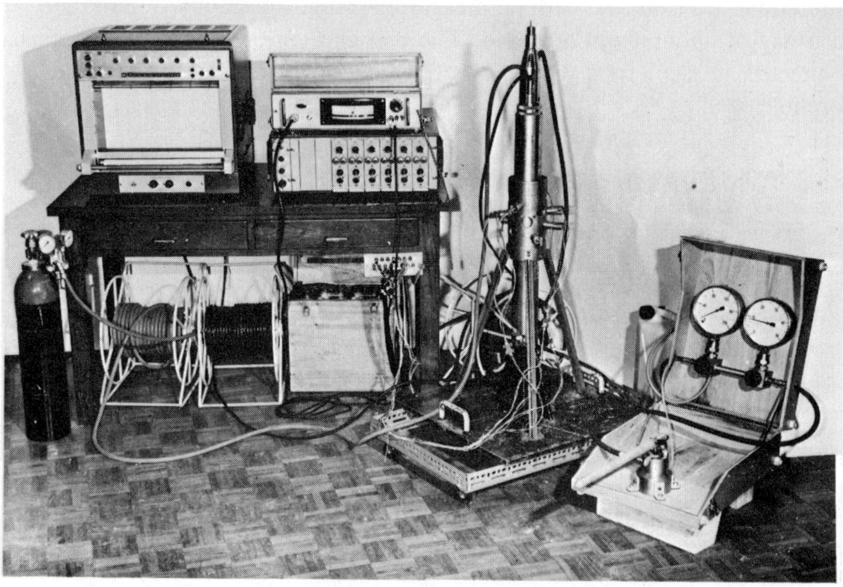

FIG. 3—*Differential transducer apparatus.*

be solved and the nature of the rock mass. It is usually advisable to carry out several loading/unloading cycles to measure the deformations for different pressure values and to observe the creep of the rock mass under the maximum pressure values.

Figure 4 shows the results obtained at a point of one of the boreholes along which deformability was determined with a view to the construction of an underground powerhouse in a limestone rock mass. The borehole is vertical and the point is located at a depth of 92 m, under 75 m of water. Deformations read by the four transducers of the dilatometer in a series of three loading/unloading cycles, during which a pressure $p = 100$ kg/cm^2 was reached, are indicated along the abscissae. The shape of the diagrams

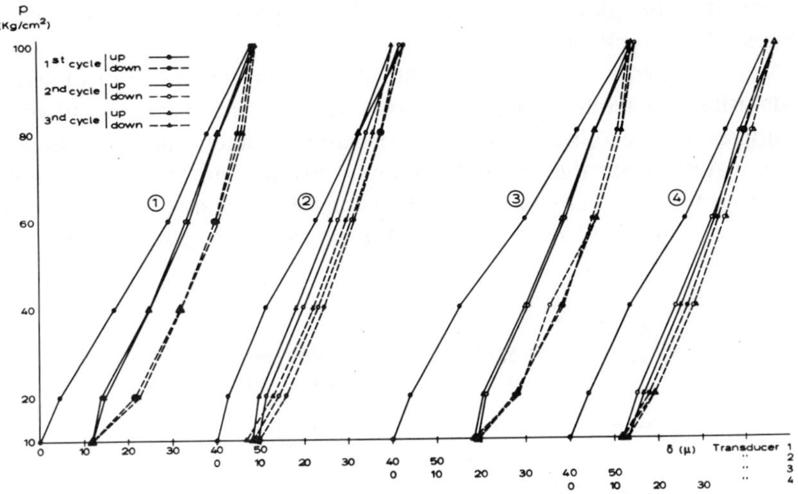

FIG. 4—*Deformations,* δ, *read by four transducers of the dilatometer for three load cycles of pressure* p.

is rather regular, and the cycles following the first are almost coincident and exhibit small permanent deformations. The deformation undergone by the borehole for the different values of the pressure can be observed in Fig. 5. The rock mass does not exhibit significant anisotropy at the point

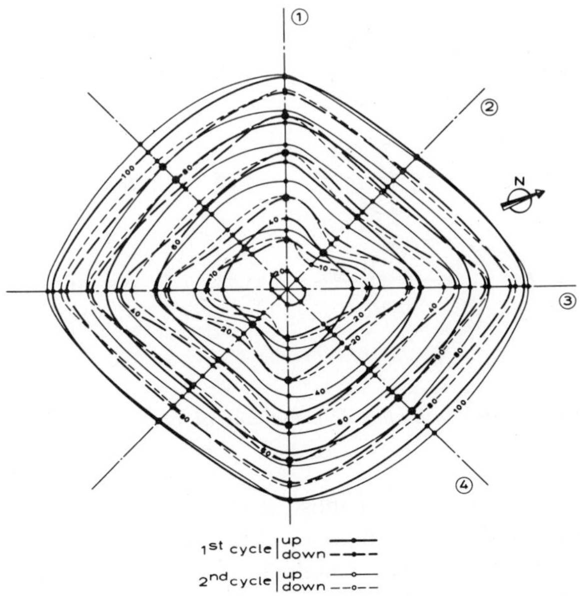

FIG. 5—*Deformation undergone by borehole for various pressures.*

observed, but the shape of the curves makes us suspect the existence of fractures in the rock mass.

Figure 6 concerns one of the boreholes tested in the course of the study of a damsite in a sedimentary rock mass consisting of limestone and shale. The ellipses of the most probable deformations (determined by the least squares method from the deformations read by the four transducers under a pressure of 100 kg/cm^2) are presented for the six points of the borehole whose depths are indicated. The position of the ellipses is related to the

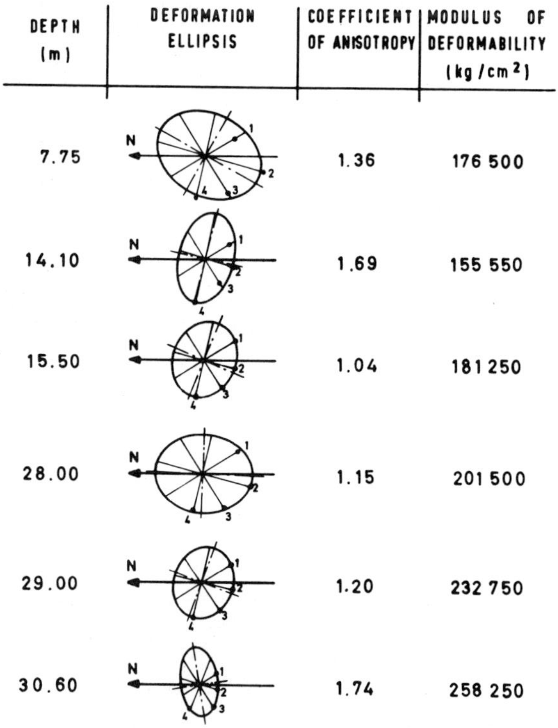

FIG. 6—*Ellipses of most probable deformations of borehole in sedimentary rock mass.*

geographic North of the site. For most of the boreholes, the values of the deformations observed show a good adjustment to an ellipses. However, as the eccentricity of the ellipses is small for most of the points where the best adjustment was found, the variation in direction of the axes observed from one point to the other should have little significance. The figure gives values for the ratio between the axes of the ellipses, which we have called the coefficient of anisotropy.

Assuming isotropic and elastic behavior for the rock mass, its Young's modulus E may be calculated by an expression of the theory of elasticity:

$$E = 2a\frac{1+\nu}{\Delta}p,$$

where $2a$ is the borehole diameter, Δ is the diameter deformation when pressure p is applied, and ν is Poisson's ratio of the rock mass. The influence of rock fractures on the deformations measured in boreholes is discussed in the work referred to above in Footnote 3.

For the test of the point at a depth of 92 m (Fig. 4), the mean value of the slope of the linear stretches of the diagrams corresponding to the second and third cycles is $\frac{p}{\Delta} = 1.2 \times 10^4$ kg/cm; hence $E = 111{,}000$ kg/cm² if $\nu = 0.2$. For the other tests referred to, the values of the modulus of deformation indicated in Fig. 6 were obtained, the value assumed for Δ being the mean value of the deformation measured in the four directions at 45 deg.

Figure 7 presents the results of tests carried out in seven boreholes, with a view toward gathering data on the deformability of a rock mass where an underground powerhouse and its tailrace tunnel will be excavated.

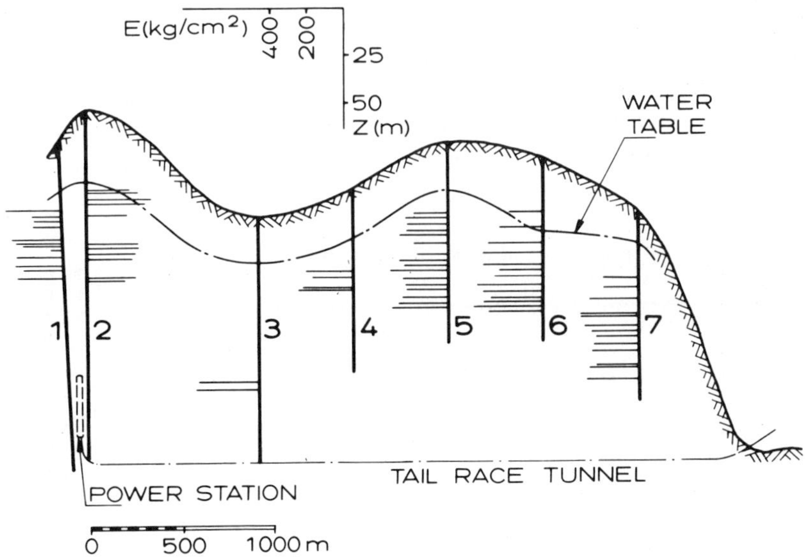

FIG. 7—*Modulus of deformation* E *for seven boreholes in rock mass above a proposed powerhouse.*

The tests extended to a depth of about 90 m, which was the limit of the dilatometer then available. The rock mass was a metasiltstone and slate interbedded with limestone, about which previous surveys had shown that its quality improved with depth. It was thus possible to conclude that the deformability of the rock mass at the depth of the planned structures is lower than that indicated by the values obtained, which were already satisfactory.

The results presented illustrate the usefulness of the dilatometer, and the experience already gained permits the assertion of its high accuracy and reliability under field conditions.

As the volume of the rock mass involved in each test is small and, therefore, often not representative of the rock mass in regards to deformability, the problem of the meaning of the results arises. Besides, the dilatometer develops a complex state of stress in the rock mass, particularly one of tensile stresses that may cause fracturing, influencing deformations in a way not yet well-known. However, since it is possible to carry out many tests and obtain an almost continuous determination of the deformability along boreholes, and since the drilling of the holes as a rule only gives rise to a minor disturbance of the rock mass, it should be possible, at least in some cases, to submit the results to statistical processing, allowing representative values of the deformability to be obtained. Having not yet enough experience to be able to form an opinion on the representativeness of the dilatometer tests, we think that the results supplied constitute at least a valuable index of deformability and, more than that, a quality index of the rock mass. It should also be stressed that there are some cases where only the dilatometer could obtain significant information about deformability, such as points deep in rock masses that cannot be reached by galleries, as is often true in the study of tunnels and underground openings or points located under river beds, etc.

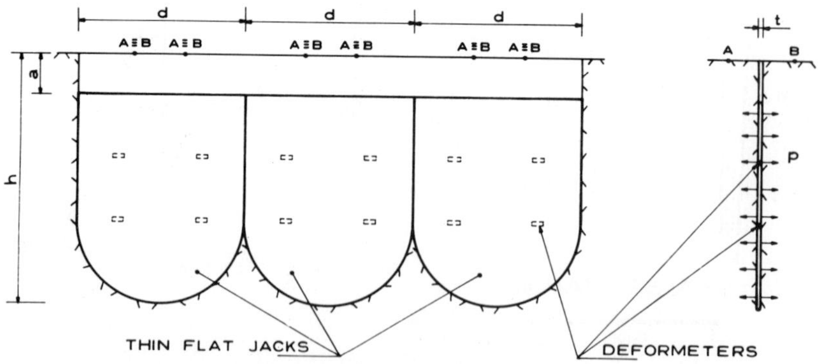

FIG. 8—*Determination of deformability by flatjack technique.*

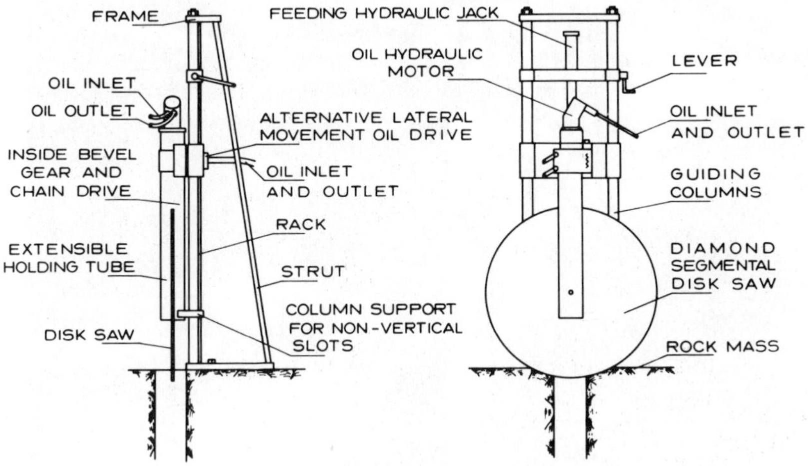

FIG. 9—*Slot-cutting machine.*

Determination of Deformability by Means of Thin Flatjacks

The technique developed essentially consists (Fig. 8) of: (1) opening with a diamond disk of diameter d a slot or a group of slots, contiguous and in the same plane perpendicular to the surface of the rock mass, with a depth h and a thickness t; (2) introducing a thin flatjack with height $h - a$ into each slot; (3) applying the same pressure p in the jacks; (4) measuring the opening δ of the slots at some points by deformeters inserted in the jacks; and (5) determination, based on (p, δ) diagrams, of the deformation moduli.

The machine shown in Fig 9 was developed to cut the slots. It consists essentially of a 1-m-diameter, 7-mm-thick diamond disk mounted at the end of a pipe, inside of which is located the system that transmits the rotating movement to the disk. The forward movement of the disk is guided by two columns belonging to the structure that allow the fixing of the machine to the rock mass. To facilitate the cutting of the rock, an alternate movement perpendicular to the forward movement is given to the disk in its plane. In order to allow the opening of slots deeper than the radius of the disk, a 17-cm-diameter hole is previously drilled by the same machine, into which is introduced the disk's supporting pipe. The machine allows the cutting of 1.5-m-deep slots, and even greater depths can be reached by using an extensive support pipe.

Figure 10 shows one of the machines and the system for operating the oil hydraulic motor, consisting of a pump moved by a petrol engine. The machine can be fixed in any position to carry out tests in the required direction.

FIG. 10—*Disk, run by hydraulically operated motor, cutting slot.*

FIG. 11—*Thin flatjack being introduced into slot cut by disk.*

Figure 11 shows a thin flatjack being introduced into a slot where two others have already been placed. Two half-cylinders are inserted into each borehole to support the flatjack. The height *a* of the unloaded area of the slots (Fig. 8) is chosen in accordance with the fracturing of the rock mass to avoid failure near the surface when pressure is applied by the jacks. If the rock mass exhibits cavities or fractures into which the steel plate of the jacks may enter, the slots should be filled with cement and the cuts repeated. The flatjacks that have been used are of the type indicated in Fig. 12. They allow the application of pressures even greater than 100 kg/cm^2.

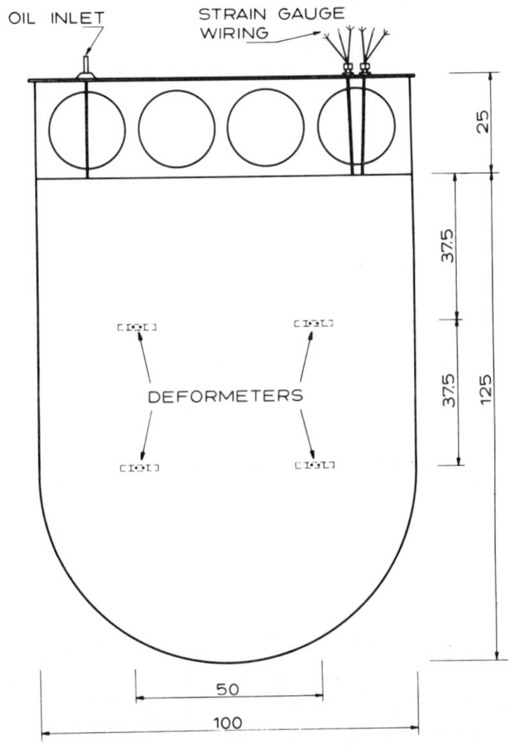

FIG. 12—*Flatjacks used in the slot tests.*

Each flatjack contains four deformeters for measuring the opening of the slots. It is thus possible to obtain a mean value of the deformability modulus and to get information about the heterogeneity of the rock mass. Each deformeter consists of a steel blade upon which electric strain gages have been mounted. The deformeters used have different measuring ranges, according to the expected opening of the slot.

The characterization of the deformability is made through a deformation

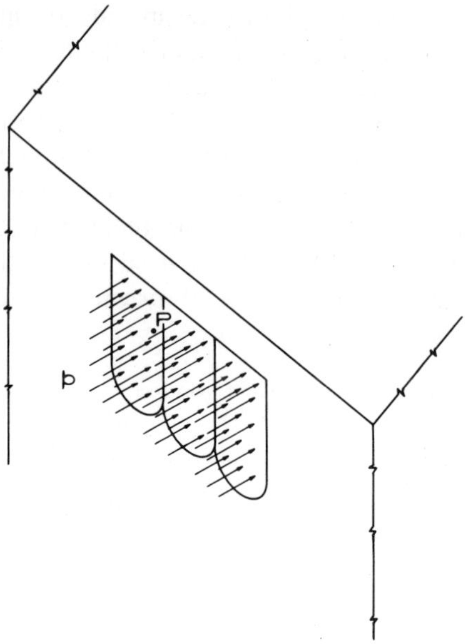

FIG. 13—*Behavior of a quarter-space under pressure.*

modulus determined by assuming an elastic behavior of the rock mass. If it is assumed that the stress concentration at the contour of the slots fractures the rock mass in the plane of the slot when pressure is applied by the flatjacks, the behavior of the rock mass may be considered as the elastic behavior of a quarter-space under pressure p (Fig. 13). The deformability modulus E at a point P of the slot where a deformeter is located is determined according to the expression:

$$E = c\frac{p}{\delta}$$

where δ is the slot opening measured and c is a constant the value of which depends on the position, shape, and size of the area under pressure p and also on the position P. This constant has the dimensions of length. Model tests on a large plaster block have been used to obtain the values of c. These tests shown an increase of about 50 percent in the displacements to the center of the loaded area when this area moves from a position far from the edge of the quarter-space (Boussinesq's solution for the half-space) to a position contiguous to that edge.

The tensile strength of the rock or its initial compressive stresses in the rock mass may limit fracturing to the extent of invalidating the representation of the equilibrium of the rock mass by the equilibrium of a quarter-space. In this case the rock mass will have to be cut by the machine developed around the area to be subjected to pressure p.

The accuracy and reliability of the thin flatjack method have been confirmed by several tests that led to entirely satisfactory results. I shall refer, in the first place, to results of tests on a very sound limestone with a high deformability modulus. These tests were carried out first in a single slot and afterwards in three contiguous slots, as is seen in Fig. 8. The characteristics of the flatjacks used are indicated in Fig. 12, where $h = 125$ cm and $a = 25$ cm. Each 1.5-m-deep slot was cut in about 4 h, despite the hardness of the rock.

Slot openings at points P_1, P_2, P_3, and P_4 observed in the single slot test are presented in Fig. 14. Two loading/unloading cycles were carried out with the jack in the position designated in the figure as "direct." The diagrams are more regular than is usual in common jackloading tests, even in very compact rock masses.

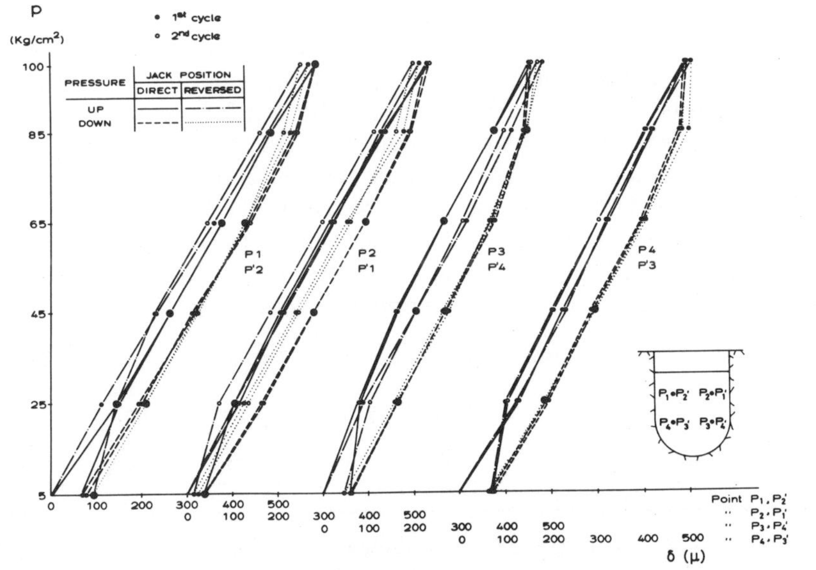

FIG. 14—*Slot openings at various pressures.*

In order to investigate the reproducibility of the results, tests were repeated after reversing the position of the flatjack. The results obtained in these tests at the points P_1', P_2', P_3', and P_4' are also indicated in Fig. 14. The diagrams corresponding to the same points of the slot show a very good reproducibility.

Deformability moduli were calculated by the expression presented above for the results obtained in each of the four positions of the deformeters. The following values corresponding to the average slopes $\frac{p}{\delta}$ of the linear stretches of the four loading and unloading diagrams presented in Fig. 14 were obtained:

Positions	c, cm	$\frac{p}{\delta}$, kg/cm	E, kg/cm²
P_1, P_2'	302	1740	525 000
P_2, P_1'	302	1740	525 000
P_3, P_4'	290	1950	565 000
P_4, P_3'	290	1950	565 000
		Mean value =	545 000

As can be seen, the values corresponding to the same distance to the edge of the quarter of space are equal, while those corresponding to a greater distance are higher, as should be expected. The values of E show an excellent agreement, which confirms, on one hand, the rock homogeneity expected from visual inspection and, on the other hand, the high reliability and accuracy of the technique developed.

After cutting three new slots in the same limestone, three types of tests were performed: tests for each one of the three slots, tests for each one of the two sets of two contiguous slots, and a test for the set of the three slots. The values of deformation modulus, in 10^3 kg/cm², obtained from (p, δ) curves according to the criterion adopted above for the single slot test are presented below; in the penultimate line the values corresponding to the measurements made with the deformeters placed at the smaller distance from the surface, that is, at points P_1 and P_2 of each flatjack, are given, and on the last line those regarding points P_3 and P_4:

One-Slot Test				Two-Slot Test			Three-Slot Test
1	2	3	Mean Value	1 and 2	2 and 3	Mean Value	
830	815	965	870	840	805	822	596
1080	915	1130	1040	890	815	852	612
	Mean value:		955			837	604

As can be seen, the values of the modulus of deformation are even greater than those referred to previously. It was found that in all the tests the values of the modulus of deformation corresponding to the readings made at the deepest points are higher. This must be due to a lower compacity of the rock mass near the surface. It should be noted, besides, that the values decrease when one changes from the one-jack tests to the tests with two jacks, and even more markedly when the test is carried out with three jacks. This fact illustrates well the influence of the magnitude of the volume tested on the value of the modulus of deformation although a very compact rock mass is being dealt with.

Time spent in a test of a single slot, including the fixing of the cutting machine, the cutting of the slot, the performance of tests according to the program indicated, and the removal of the jack from the slot, is only about 10 h. In the case of three-slot tests the time needed is about 3 days.

The test program to study the deformability of a rock mass depends on the problem to be solved and on the nature of the rock mass. First, the total area of the slot depends on both the degree of heterogeneity and the spacing of fracturing of the rock mass. In applying the pressure p the following procedure is advisable: perform several loading/unloading cycles, measure the slot openings for successive values of the pressure, and observe the creep of the rock mass for the maximum value of the pressure and maybe for other values. It is also necessary to observe the evolution, during the tests, of the crack that usually occurs in the plane of the thin flatjacks.

To investigate the influence of grouting, tests can be repeated in the same slots after grouting the rock mass under conditions similar to those occurring in the structure. As previously mentioned, the location of the test sites should be chosen in accordance with the zonation of the rock mass, which will be based on the results obtained with the dilatometer.

From what has been said it can be concluded that thin flatjacks allow the quick and economical testing of large volumes of practically undisturbed rock mass. In fact, the disturbance induced by the cut of the slots is negligible, and the disturbance resulting from the preparation of the rock surface is very small at depths with influence on the slot opening and in the direction of the pressure applied by the jacks. Note that in current jack-loading tests, pressures are applied in the direction where decompression due to the preparation of the surface is more marked.

With this new technique the difficulty usually met with in the determination of the deformability of a rock mass, particularly insofar as foundation problems are concerned, is overcome. A paper in which the technique and its applications are dealt with more fully is presently being prepared.

Correlation of Results

Correlations between results obtained with the dilatometer and with flatjacks are of much interest. In fact, these correlations will allow a better understanding of the meaning of the results supplied by the dilatometer, a reduction of the number of flatjack tests, which are more time-consuming and expensive than dilatometer tests, and the application of the results obtained with flatjack tests to the zones tested only with the dilatometer.

The easiest way to obtain these correlations is to use the cutting machine itself to drill a 76-mm borehole and to perform the dilatometer tests, preferably at the depths at which the deformeters of the flatjacks will be placed. Afterwards, the 17-cm hole will be drilled and the slot cut and tested, the cutting machine always being kept in the same position.

If correlation with results of laboratory tests on rock specimens is to be made, it is not advisable to drill the 76-mm holes in the position indicated to allow tests of larger specimens taken from the 17-cm drilling cores. The testing of specimens taken along the direction in which pressure is applied by the jacks is particularly relevant. The dilatometer tests can then be performed in holes drilled near the slots; for instance, in the four holes indicated in Fig. 15 for a three-flatjack test.

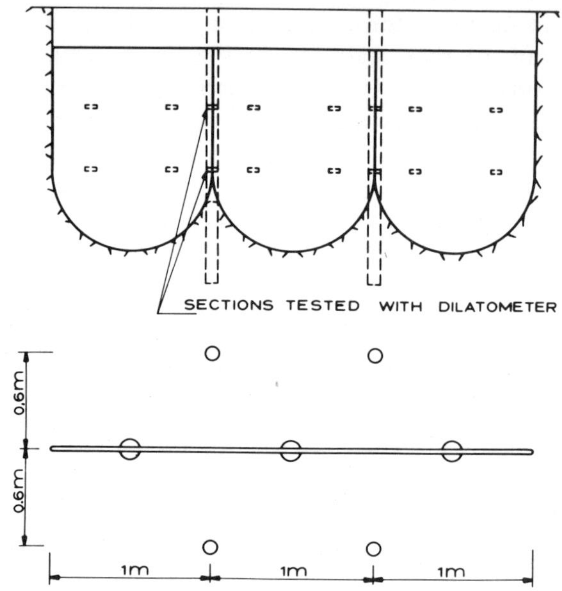

FIG. 15—*Configuration of three-flatjack test.*

We are determining these correlations whenever tests with flatjacks are performed. In conclusion I would stress that the two techniques described, used jointly, allow, in most cases, the characterization in satisfactory terms of the deformability of rock masses, which so far was not usually possible.

DISCUSSION

O. S. Cecil[1] *(written discussion)*—Dr. Rocha presented results from dilatometer tests that were conducted in vertical drill holes above a horizontal tailrace tunnel in Brazil. For what purpose were these tests conducted, and how, specifically, was the information used in design?

Also, for what types of projects are the dilatometer and flatjack modulus tests expected to be used? Sketches of civil engineering structures and the proposed test setups would be helpful in explaining this question.

M. Rocha (author's closure)—As indicated in the final version of my paper, the purpose of the tests was to investigate the change of deformability with depth, with a view toward correlating the test results with the geologic data yielded by the cores sampled in the boreholes. It was concluded that the rock mass has a rather high modulus of elasticity at the depth planned for the powerhouse and the tailrace tunnel.

As I had occasion to emphasize, the use of the dilatometer will make it possible to obtain, preferably during the geologic exploration, an estimate of the deformability of the rock mass; on the basis of this estimate the rock mass can be zoned and the deformability of each zone can be subsequently determined by the flatjack method. In the case of deep tunnels or other deep openings to which the access to the zones where the structure will be built is difficult, tests with thin flatjacks can be performed in zones nearer the surface considered (on the basis, especially, of dilatometer tests) as representative of the deep zones; these zones can then be tested when they are made accessible during the construction. Additionally, as more experience is gathered with the use of dilatometers, it will be possible to use known correlations applicable in the case under consideration.

W. S. Brown[2] *(written discussion)*—In his very interesting and informative paper Dr. Rocha presented deformation modulus results[3] from thin

[1] Missouri River Division Laboratory, Corps of Engineers, U.S. Army, Omaha, Neb. 68102.

[2] Head, Structural and Rock Mechanics Section, Division of Research, U.S. Bureau of Reclamation, Denver, Colo. 80225.

[3] Rocha, M., "New Techniques for the Determination of the Deformability and State of Stress in Rock Masses," presented at International Symposium on Rock Mechanics, Madrid, 1968.

flatjack tests (p. 52). The modulus results vary with the number and the position of the jacks. The author attributes the lower modulus near the surface to a lower capacity of the rock mass in that location. The trend of the data with position indicates that the stress field and edge restraint may be a more significant reason for the low modulus values near the surface.

The test results show that data obtained near the edges of the slot (both sides and bottom) always result in higher modulus values. This is due to the restraint provided by the rock mass and indicates that a longer and deeper slot is necessary to develop a plane stress field. It is not clear that the author's calibration of the thin flatjack could account for this problem, since apparently all jacks were calibrated in the same calibration block.

Examination of the modulus data given above shows that when loaded one at a time the center jack gave a lower modulus than the outside jacks. This should be expected since the restraint provided by the slot is higher at the edges. This is verified by the two-slot tests where use of the center jack with either of the outside jacks gave reasonable agreement.

It is interesting that the modulus values decreased as the number of jacks in a test increased. However, it is disconcerting to see that the incremental difference in modulus becomes larger rather than smaller as the number of jacks is increased. One would certainly expect to reach an asymptotic value and the limited results from three tests does not indicate such a trend. It would be worthwhile to extend the test results with several more flatjacks to insure that the results approach an asymptotic value. This may be helpful in determining how many jacks are really necessary to give reliable results.

M. Rocha (author's closure)—Professor Brown has raised a very pertinent question, namely, to what extent does fracturing of the rock in the plane of the thin flatjacks invalidate the representation of the equilibrium by that of a quarter-space. In fact applications of the method after the Denver symposium have shown that sometimes insufficiently developed fracturing that depends on the tensile strength of the rock mass and on its initial state of stress is observed. Meanwhile, the influence of the extent of the development of fracturing on the moduli of elasticity obtained was investigated by analytical means.

In the case of the results under consideration an influence of the edge restraint is quite possible. Nevertheless, as regards the lower moduli yielded by the deformeters placed nearer the surface, it should be noted that a certain weathering of the rock mass near the surface was disclosed by the core samples.

Regarding the other comment, namely, the decreasing value of the modulus of elasticity for larger volumes of the rock mass influenced by the test, I think that this marked decrease of the modulus when the number of flatjacks increased from two to three is due to the presence of a joint at a certain distance from the slots. As for the number of flatjacks required to obtain a representative value of the deformation, there is no general rule. It depends on the accuracy required for the value of the deformation, or the heretogeneity of the rock mass, and on the distance between joints and other fracturing. Thus each case has to be assessed separately.

G. H. Kruse[1]

Deformability of Rock Structures, California State Water Project

REFERENCE: Kruse, G. H., **"Deformability of Rock Structures, California State Water Project,"** *Determination of the* In Situ *Modulus of Deformation of Rock, ASTM STP 477,* American Society for Testing and Materials, 1970, pp. 58–88.

ABSTRACT: An anlysis of the convergence of rock that accompanies the driving of a tunnel or underground chamber is presented. The analysis was used to augment conventional methods of determining *in situ* modulus of deformation for the design of an underground power plant chamber. An evaluation of rock behavior in the chamber is presented in which the relaxation analysis is used. Use of the relaxation analysis in determining *in situ* modulus of deformation for the design of the steel liner of a high-pressure discharge line in rock is also discussed.

KEY WORDS: rocks, rock mechanics, strains, subsurface structures, modulus of elasticity, determination of stress, deformation, electric power plant, pressure tunneling, evaluation

It is logical to assume that the deformations which accompany the driving of a rock tunnel or underground chamber are related to the same parameters which affect other *in situ* rock deformations, that is the deformation modulus of the rock and the geometry of stress variations. In the past six years the California Department of Water Resources has made several hundred measurements of tunnel convergence, or "relaxation," in two of the principal underground features of the State Water Project. These measurements, and the analytical methods used in evaluating them, provide the subject matter for this paper.

The State Water Project is a water conservation and conveyance system over 500 miles in length, which links the water-surplus area of northern California with the water-deficient areas of southern California. It was begun in the 1950's and will be completed in the early 1970's at a cost of over two billion dollars. Initial construction activity was centered in

[1] Associate engineer, California Department of Water Resources, Sacramento, Calif.

northern California on the construction of Oroville Dam and its appurtenant features, including an underground power plant. Construction activity later shifted to the southern part of the state, where many miles of tunnels are being constructed to convey water from the Central Valley of California on its southward course through the Tehachapi Mountains. The major pumping lift of this conveyance system is a single 1900-ft lift with twin discharge line tunnels designed to utilize the natural rock as a pressure-bearing material. Initially, the underground power plant chamber, and later, the twin discharge lines, were subjected to extensive *in situ* rock testing for evaluations of structural behavior. The term "rock structures" used in the title of this paper is intended to describe rock masses which are designed to bear loads in excess of normal gravity, thereby supplanting extra concrete or steel which would otherwise be required at a greatly increased expense.

Determination of *In Situ* Modulus of Deformation

Webster[2] defines the word "modulus" to be "a real positive quantity, numerical or physical, that expresses the measure of some function, property or effect . . . under unit conditions . . ." In dealing with a homogeneous, isotropic, and elastic material within its elastic limit, the modulus of elasticity, or Young's modulus, is ideally a constant considered to be a discreet property of the material. The application of elastic theory to rock, including all of its joints, fractures, and schistosity, has been found to deviate from this ideal. In recognition of this fact it has become accepted practice to place increasing reliance upon the testing of rock in its natural state so that the modulus thus determined reflects the composite action of both the rock and its defects. Such procedures require tests of the largest possible scale so that the greatest number of defects, which are often widely dispersed, will be included in the testing.

To determine the *in situ* modulus of deformation of a rock mass, two requirements must be met: (1) the rock deformation and the stress increment producing it must be measurable and (2) an analytic method of describing the geometry of the stress-deformation relationship must be developed. The latter relationship is usually evolved from the linear stress-strain relationship of elastic theory. Thus, the Boussinesq equations describe the deformation of a semi-infinite solid under applied load and provide the method of computing rock modulus from plate bearing or jacking tests [*1*].[3] Analyses for thick-walled cylinders are utilized in a similar manner when hydrostatic pressures are involved [*2*]. The analysis of the effect of a hole in a plate has been applied to evaluations of borehole relief tests and flatjack tests [*3, 4*].

[2] New Collegiate Dictionary, Fifth Edition, 1936.

[3] Italic numbers in brackets refer to the list of references at the end of this paper.

All of these are applications of elastic theory using simple mathematical models. Much of what has been learned about rock behavior and utilized in the design of rock support for civil engineering structures has been derived from such analyses. Applications of more sophisticated types of mathematical analyses, such as the finite element method discussed later in this paper, utilize the flexibility of electronic computers for dealing with geometric models for which there are no simple mathematical representations. The geometry of the model is approximated by a network of elements. Elastic constants assigned uniformly to all elements cause the model to respond elastically, like the simple mathematical models.

A further capability of such a method of analysis is the simulation of nonelastic behavior through the use of incremental variables for elastic "constants." Individual elements of the model are made to respond in compliance with elastic "constants" which are varied in some predetermined manner. Because such analyses presume a prior knowledge of the properties being tested, they do not lend themselves to calculation of those properties from *in situ* test data. The calculation of an *in situ* modulus of deformation, therefore, is usually based upon the assumption of elastic behavior, because the alternative is to introduce other assumptions about variables (constants) which cannot be readily substantiated even from reasonably thorough exploration of the geologic fabric of the rock mass. This, of course, does not preclude the need to evaluate elastic analyses with informed judgment attained through a thorough examination of the geology of the test site and to make such modifications as may be warranted by the data available.

Although it is customary for papers on rock mechanics to list references on elasticity, there has never been published in the United States a general mathematical analysis of elastic deformation analogous to the relaxations which accompany the driving of a tunnel. This is in contrast to numerous publications which portray analyses of stress concentrations around tunnels based either upon the Kirsch equations for circular openings, or the photoelastic, or Moiré analyses for circular and other shapes.

The earliest publication of an analytical method of evaluating radial tunnel deflections was reported by Alexander [5] in Australia. It was derived from Timoshenko's development for the deflections in a plate with an elliptical hole. Alexander transformed the awkward curvilinear coordinates used by Timoshenko into Cartesian coordinates. The transformation, however, destroyed the general nature of the generic equation and restricted its application with respect to directions of stress. As originally reported, the equation was applied to relaxations in the 14-ft diameter Tooma-Tumut tunnel, which were monitored by extensions of hollow bolts placed in 8-ft boreholes drilled radially from the tunnel axis. The deformation modulus of the rock envelope was calculated from rock stresses determined by flatjack measurements. Alexander reported that bolts installed 4 ft from the heading produced displacements "so small that the rock must be considered undisturbed by the tunnel advance. At one foot from the heading, displacements were large."[4]

Although instruments to measure radial tunnel deflections have been reported in the United States [*5, 6, 7*], no efforts to relate deflections to tunnel stresses have been reported. This may be attributed in part to the use of instruments too susceptible to blast damage to be installed near enough to tunnel headings to obtain adequate measurements of the full relaxation. Borehole extensometers are often installed in areas where the initial relaxations of excavation have already occurred. They are useful in monitoring long-term creep or stability, but their results are not applicable to the evaluation of deformation modulus.

Displacements Around an Opening

The derivation of equations for relaxation displacements around rock openings is based upon the following assumptions: (1) the displacements are dependent upon the direction and magnitude of the primary stresses existing in the rock and (2) the relaxations normal to the axis of the opening are analogous to the displacement in a thin plate subjected to biaxial stress.

Two equations that describe the displacement in a plate are used:[5] the first gives the displacement in a plate with a hole,

$$u = u\,(E,\ a,\ r,\ \theta,\ \nu) \qquad (1)$$

while the second gives the displacement in a continuous plate,

$$u_0 = u_0\,(E,\ r,\ \theta,\ \nu) \qquad (2)$$

The difference between these equations gives the displacement induced by the hole:

$$u_x = u - u_0$$

It is the displacement that would occur if the hole were formed in an already stressed material. The equation for the radial relaxation around a circular opening is

$$u_x = \frac{Sa^2}{2Er}\left[(1+\nu)\,(1 - \frac{a^2}{r^2}\cos 2\theta + 4\,(1-\nu^2)\cos 2\theta\right] + \frac{Qa^2}{2Er}\left[(1+\nu)\,(1 + \frac{a^2}{r^2}\cos 2\theta) - 4\,(1-\nu^2)\cos 2\theta\right] \qquad (3)$$

Graphical solutions to Eq 3 are shown in Fig. 1 for four different ratios of biaxial stress and with Poisson's ratio fixed at 0.25. The relatively minor

[4] Page 167.

[5] Analytical functions and nomenclature are shown in the Appendix.

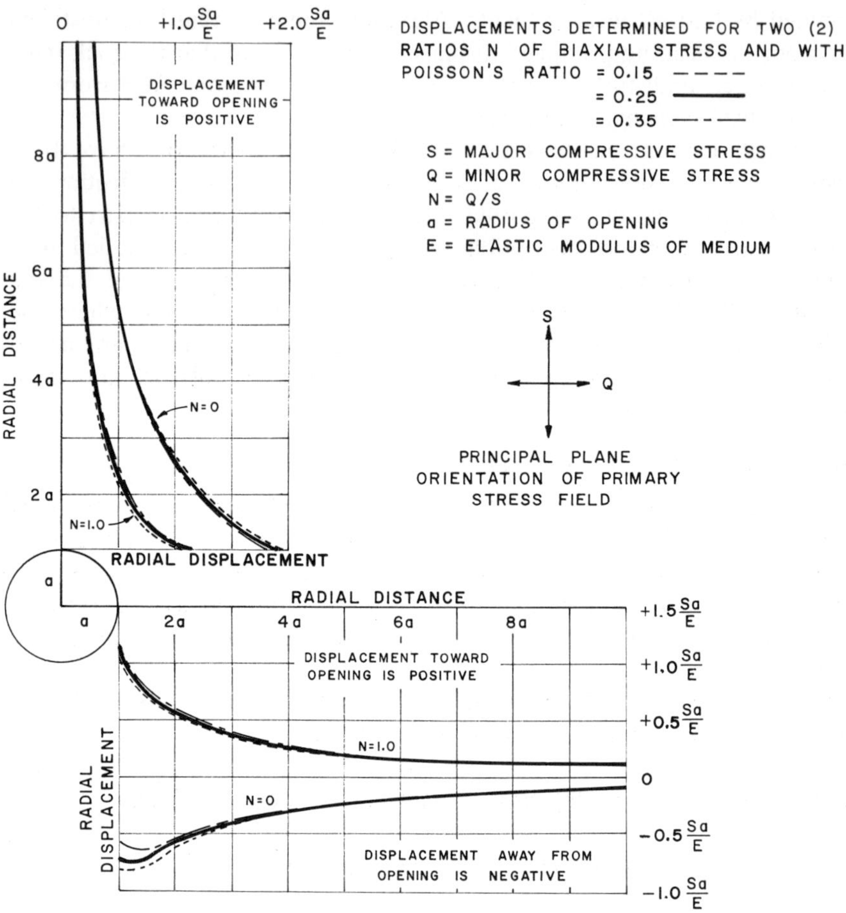

FIG. 1—*Displacements in a prestressed medium induced by the extraction of a circular section.*

effect of Poisson's ratio is illustrated in Fig. 2. A similar development for an elliptical hole is derived from Eq 4 in the Appendix. Graphical solutions for an elliptical hole are shown in Figs. 3 and 4.

In a similar way, analyses by the finite element method may also be used to determine these displacements, since it is especially useful for irregularly shaped openings. An element network which defines the boundary of the opening but is continuous over its area is drawn. Two computer runs with appropriate inputs to describe the deflections with and without the hole are made, and the differences are plotted.

A finite element analysis of relaxation for the crown opening of the chamber for the Oroville power plant and an analysis using an elliptical approximation of the opening are compared in Fig. 5. Radial relaxations

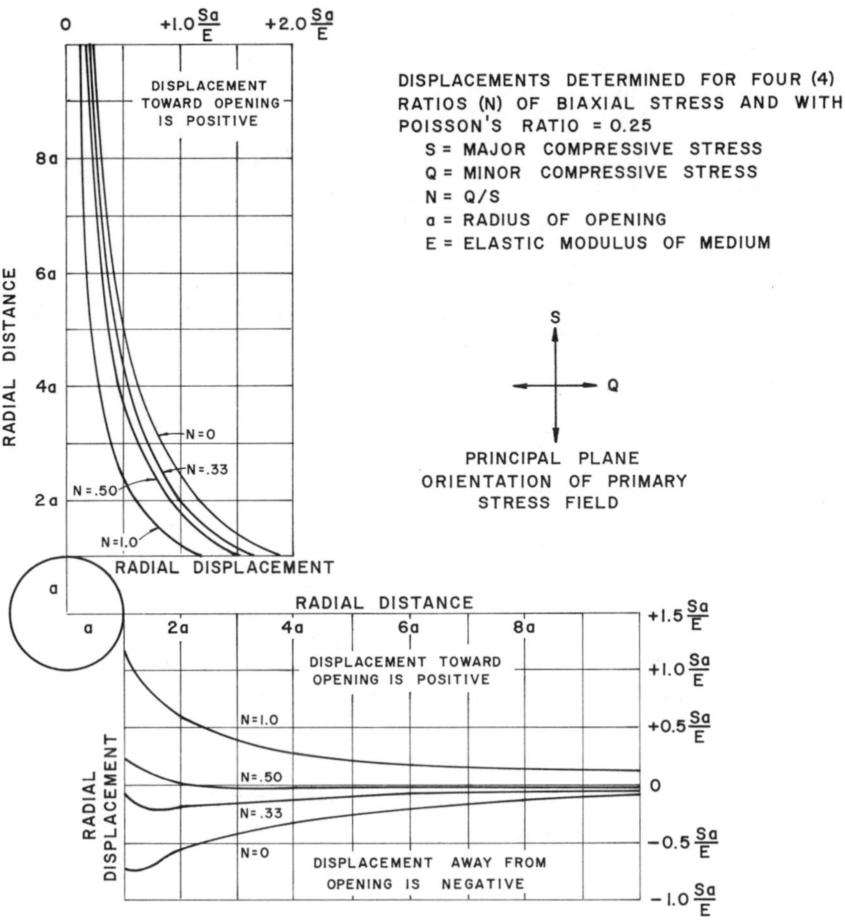

FIG. 2—*Displacements in a prestressed medium induced by the extraction of a circular section.*

at the center line are shown to be closely comparable for either method. When a reasonably good approximation of an opening can be made with an elliptical hole, the analysis is simplified since no element network need be drawn.

Determination of Rock Modulus from Relaxations—Initial efforts to determine rock modulus from relaxations were made in 1962 during the construction of the first of two 35-ft-diameter diversion tunnels at the site of Oroville Dam. The objective was to supplement the more conventional methods of modulus determination then being used to evaluate the performance of rock in the design of a large chamber for an underground power plant. The primary stress field in the rock mass had already been evaluated from flatjack tests and borehole stress relief tests [*8, 9*] per-

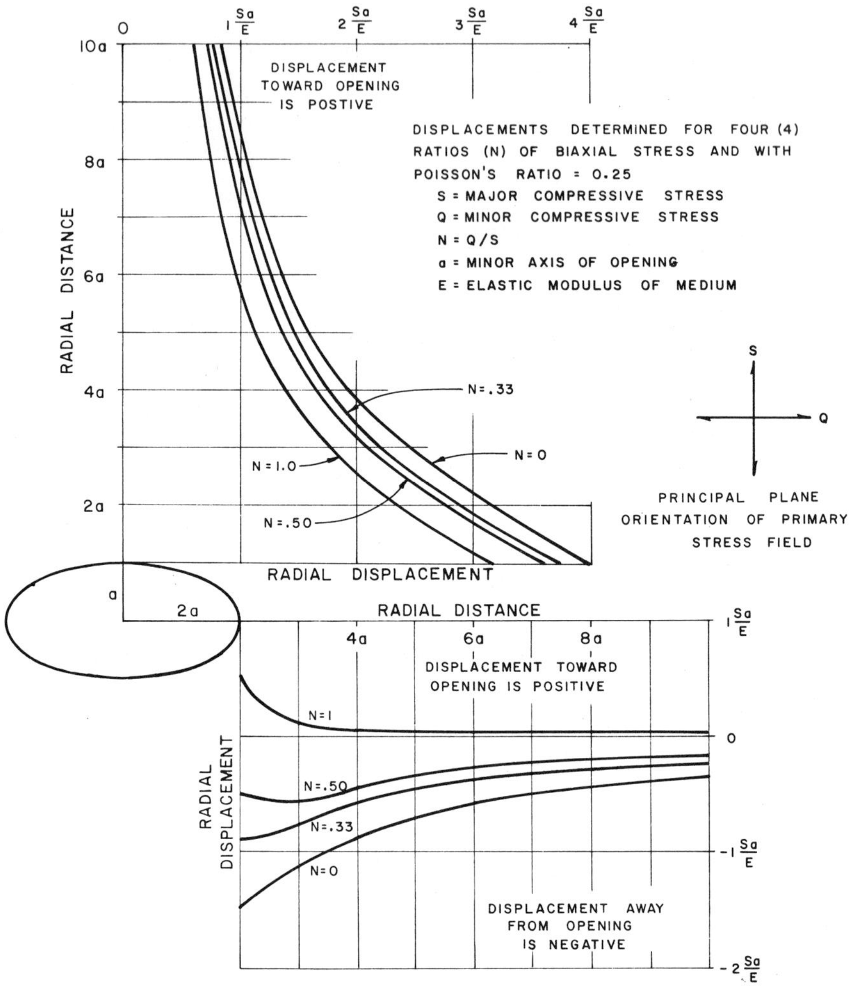

FIG. 3—*Displacements in a prestressed medium induced by the extraction of a typical elliptical section.*

formed in an adjacent exploration tunnel. Twenty-two extensometers were installed to measure the relaxations caused by the advancing heading. To avoid slowing the contractors 6-day weekly rate of progress, a radial ring of four or five extensometers was installed each Sunday by Department personnel as close to the heading as drilling would permit. Consecutive rings were thus established approximately 100 ft apart. The single-span extensometers were rock bolts adapted for extension measurements. Relaxations were measured by dial gages which recorded to within 0.001 in.

The tunnel was excavated in a hard dense rock of specific gravity over

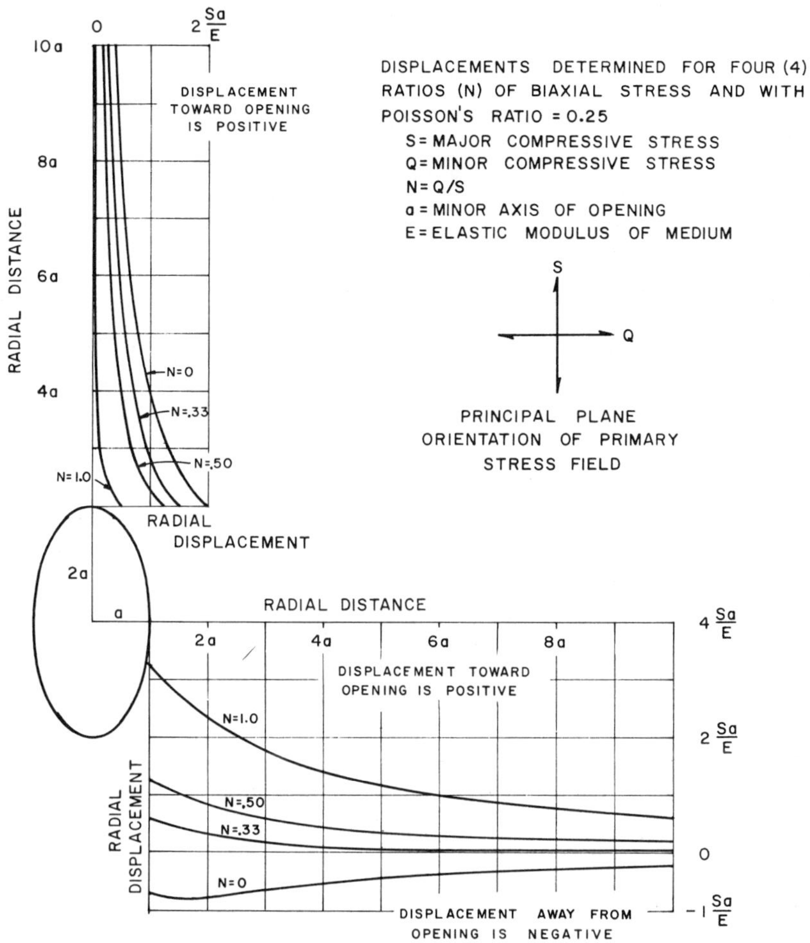

FIG. 4—*Displacements in a prestressed medium induced by the extraction of a typical elliptical section.*

2.9. Since the rock is metamorphic but contains high amounts of amphibole, it is best described as amphibolitic. It appears massive in spite of a common, rough schistose structure. Narrow bands of schist and a system of well-healed joints and occasional shear zones are found throughout the test area.

Relaxations measured in the diversion tunnel were used to calculate deformation moduli using the measured uniform stress field of 500 psi. The results are shown in Table 1, along with results from other types of tests made in exploration adits preliminary to the construction of the power plant chamber.

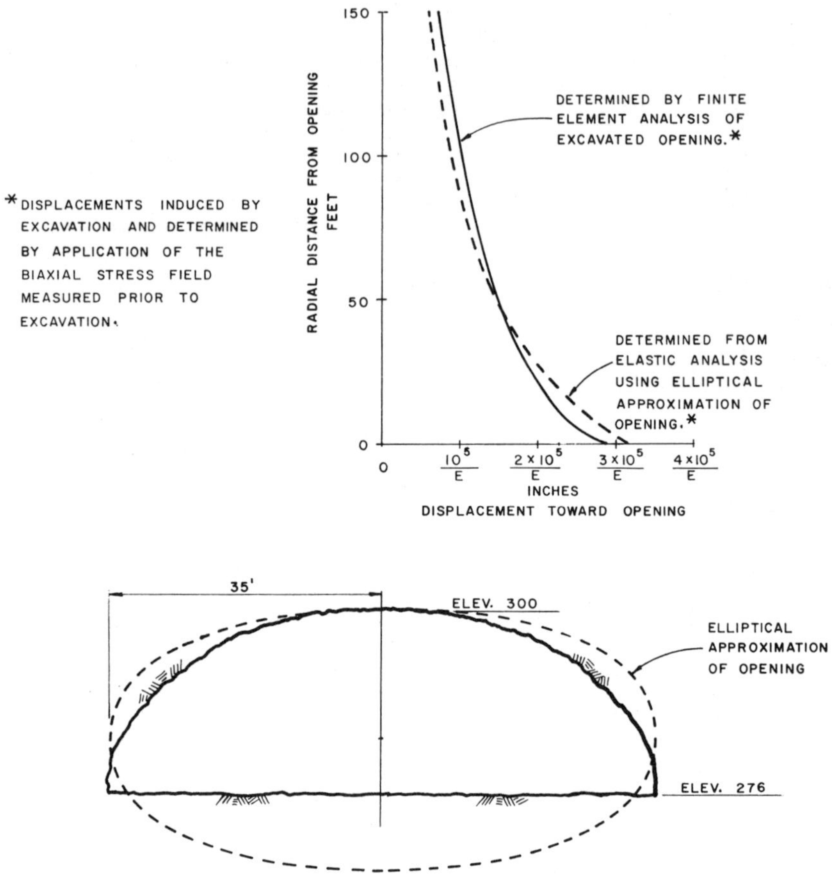

FIG. 5—*Analyses of displacements for crown opening—Oroville underground power plant.*

TABLE 1—*Summary of rock moduli measured at Oroville damsite.*

Type of Test	Number of Measurements	Rock Modulus, psi Range	Average
Static tests on cores	21	10.8 – 15.2 × 10^6	12.9 × 10^6
Sonic tests on cores	3	14.8 – 17.0	16.0
Geophysical-seismic velocity	19	4.9 – 15.5	10.2
Flatjack	30	1.4 – 16.4	7.5
Tunnel relaxation	22	0.6 – 7.5	2.6
Plate bearing	5	1.2 – 1.8	1.5

Comparison With Rock Modulus Determined by Other Methods

Probably no large underground chamber has ever been constructed for which more different types of prelminary tests for *in situ* rock properties were made. A comparison of the different tests and of the wide range of values obtained is inclined to be somewhat speculative but, nevertheless, informative.

The scatter of flatjack results was unexpected, because the wide differences were in no way discernible from a visual inspection of the sites. In only one case was it possible to attribute a low modulus to the possibility of joint action in the vicinity of the jack. In fact the flatjack sites had all been chosen very carefully for the visual quality and apparent massiveness of the rock.

The scatter of relaxation measurements was more readily understood since these sites were established by the chance of construction progress. Relaxation measurements are cumulative deformations of blocks and movements of joints. Blocks may move in translation but may also rotate slightly. An individual measurement may depend on what part of a given block the extensometer is anchored.

Of greater significance than the scatter of individual measurements was the lack of agreement found in the average rock modulus computed for each of the three types of tests. These differences suggest that the modulus may be a property of the test as well as of the rock, and that the distinct values of these averages result from the geometry of the test relative to the conditioning of the rock by the tunnel relaxations. This is more fully understood after considering the unique characteristics of the various test procedures.

Flatjack Tests—The primary purpose of a flatjack test is stress measurement. After installing a pattern of reference pins in the rock wall or roof of an exploratory adit, initial measurements are made between pins. When a slot is cut into the wall, which induces a stress relief, the pins on opposite sides of the slot converge, and measurements between the pins are repeated. Relaxations of the slot are thus recorded. A steel flatjack is then grouted into the slot and inflated with fluid under pressure until the reference pins are returned to their original distance of separation. This procedure is called cancellation, and the pressure required is assumed to be the original stress in the rock.[6]

The analysis of the flatjack slot is made by simulating a cross section of the slot with a slender, elongated elliptical hole. The analysis is essentially a hole-in-a-plate relaxation analysis. The original relaxation of the drilled slot, along with the stress determined from cancellation, provides a basis for calculating the rock modulus.

It should be noted that flatjacks are installed so that the jacking pressure is opposed by either a tangential or a longitudinal rock stress. Since neither of these stresses is relieved or diminished in the driving of the tunnel, the

[6] The hazards of this assumption are discussed in Ref *9*.

rock structure along these orientations remains more nearly in its original condition of soundness than does the rock structure in the radial direction. A relaxation anisotropy is a natural consequence of tunnel excavation. For this reason, the modulus determined from the flatjacks (7.5×10^6 psi average) probably represented an upper limit for the rock modulus in the Oroville chamber, applicable only to areas where relaxations could be greatly restrained.

Plate Bearing Tests—While the geometry of flatjacks prohibits load applications and deformation radial to the tunnel axis, this is not a limitation of plate bearing tests [*10*]. Jacking loads applied across the tunnel diameter produce radial deformations at the tunnel walls. A consequence of relaxation anisotropy during the driving of a tunnel is that radial stresses diminish to zero at the edge of the opening, while tangential stresses become concentrated. Joints or fractures which have nearly tangential orientations have the greatest tendency to open. This condition influences the borehole extensometer measurements as well as the magnitude of compressions generated by plate bearing loads. It is therefore not surprising that the rock moduli computed from radial deflections would be lower than those determined from flatjack tests.

The average modulus calculated from relaxations was 2.6×10^6 psi, and from plate-bearing tests, 1.5×10^6 psi. Although these figures are relatively divergent when considered alone, they are reasonably close when compared with results of other tests shown in Table 1. For purposes of evaluating behavior in the large chamber, either value would have been acceptable although, as is the custom in design, the most conservative (lowest) value was applied in performance analyses.

Rock Stresses

In analyzing the power plant chamber for purposes of design, considerable attention was given to the stresses around the opening. Stress analyses by the finite element method were carried out at the University of California under the direction of J. M. Raphael and R. W. Clough, Jr. Vertical and lateral loads on the chamber were adapted from *in situ* measurements of stress. Analyses were made for full and partial openings. A typical result of the stress analysis is shown in Fig. 6, which portrays contours of equal minimum principal stress. These stresses were essentially tangential about the opening.

Further borehole stress relief tests were carried out at the center line of the arch to verify and evaluate the stresses predicted from the finite element analysis (results are shown in Fig. 7). Although the measurements were in reasonably good agreement with predicted stresses at depths beyond 4 or 5 ft, they were somewhat greater at shallower depths. The reason for this is attributed to the reinforcement pattern of rock bolts that had

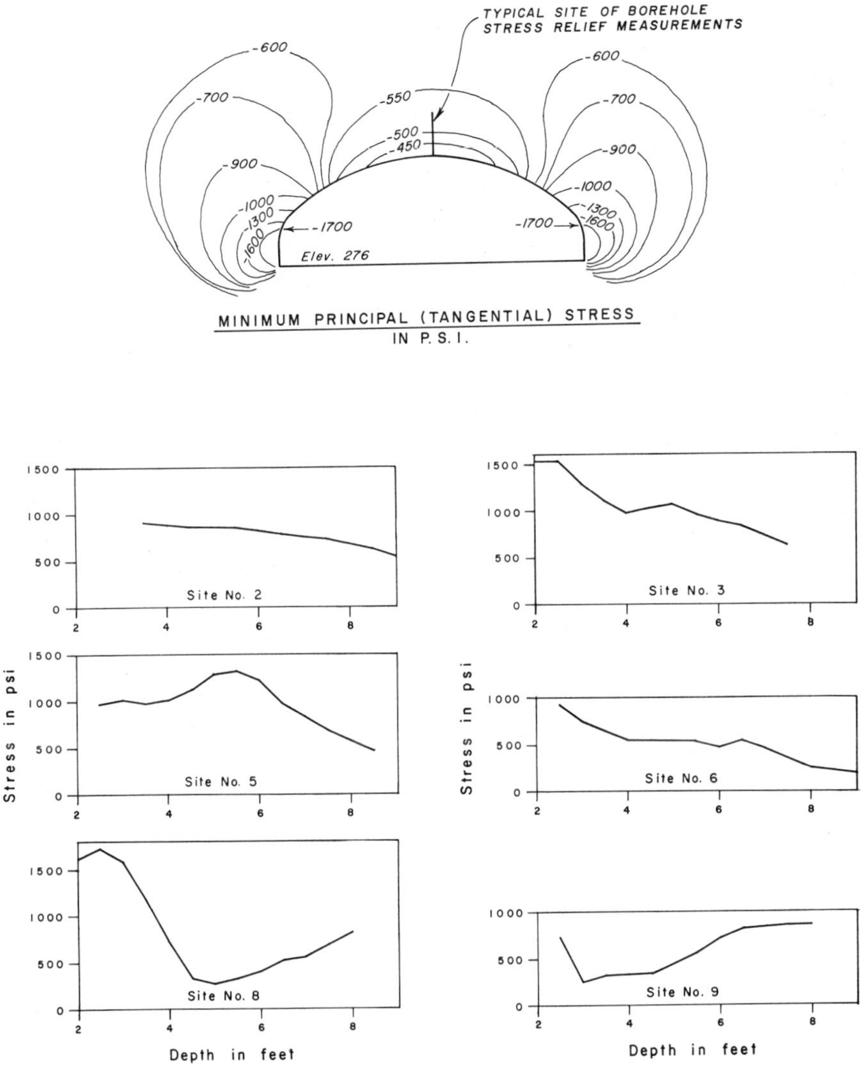

FIG. 7—*Borehole stress relief measurements of tangential stress, center line of rock arch—Oroville underground power plant.*

been installed on 4-ft centers. Each stress relief borehole was necessarily bounded by four rock bolt bearing plates, with some as close as 1 ft from the collar of the borehole. Although the bolt-applied stresses do not directly account for the magnitude of stress increase, rock confinement near the bearing plates is believed to stiffen the rock enough to produce localized secondary concentrations of tangential rock stress capable of being detected by the stress relief measurements.

Extensometers for Power Plant Chamber

Relaxations in the crown opening of the power plant chamber were monitored by 20-ft mechanical borehole extensometers which were modifications of those used earlier in the diversion tunnel. They were single-span instruments in which the deep anchor was a standard bolt anchor welded to a ¼-in. diameter steel rod. This anchor was set in the hole and tightened with a socket wrench mounted on a 21-ft pipe extension. The tip of the rod was then passed into a short section of hollow rock bolt anchored at the collar of the borehole. Stainless steel reference surfaces were provided for dial gage measurement. The instrument was designed for fast installation directly from the drilling jumbo while rock bolting was in progress. With experience, installation time was shortened to as little as 10 min. Thirty-nine extensometers were installed along the 540-ft length of the chamber on or near the center line of the arch. These mechanical instuments were monitored regularly until superseded by 40-ft single-span extensometers equipped with electrical strain gage transducers for remote monitoring. During the deepening of the chamber by benching, the rock was monitored by remote-reading extensometers installed in both arch and walls.

It is perhaps paradoxical that the magnitudes of relaxation recorded by these instruments was of secondary importance. The primary purpose of the instruments was to reveal the character of the relaxations as an aid for evaluating stability. Time-deformation graphs reveal the character of deformation and permit detection of abnormal creep following the initial excavation. They are initiated with the first stress relief of excavation and

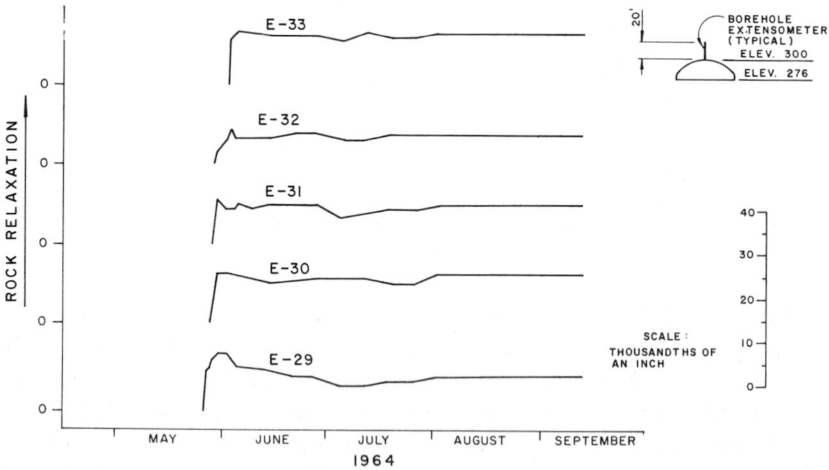

FIG. 8—*Arch deformation versus time—Oroville underground power plant.*

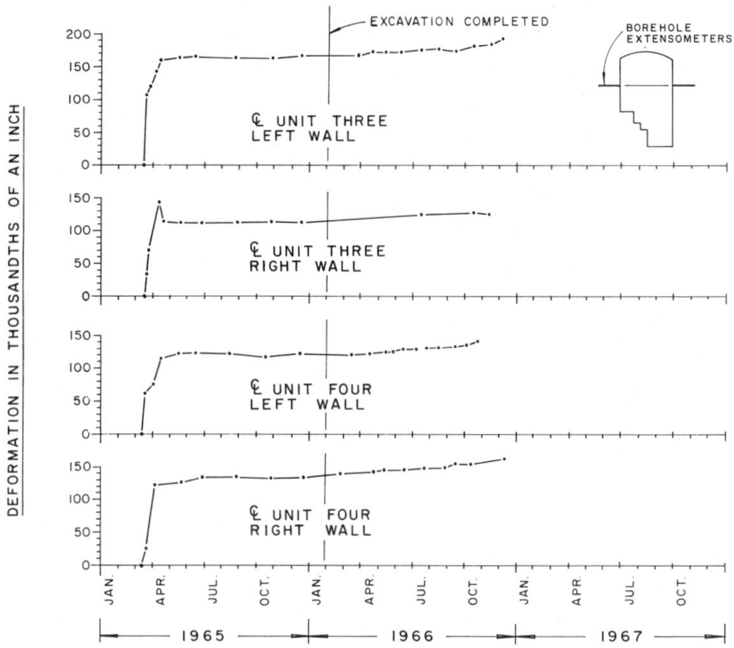

FIG. 9—*Wall deformation versus time—Oroville underground power plant.*

are continuously updated for creep analysis. Typical graphs are shown in Fig. 8 for the crown opening and in Fig. 9 for the full opening. It is an indication of the high quality of the foundation rock in the area of the underground chamber that no measurable evidence of instability was found, either during the excavation or subsequently. Relaxations measured by the borehole extensometers were about equal to, or in many cases a good deal less than, what was expected on the basis of preconstruction rock testing.

The initial excavation of the crown opening was a critical phase of construction. Specifications called for rock reinforcement to be installed on 4-ft centers. Bolting of the rock was required to be completed to within 5 ft of the new heading within 3 h after each blast. Prior to the excavation of this chamber, none of the tunnels in the Oroville underground complex had been supported by rock reinforcement. Location of the 20-ft borehole extensometers was at the discretion of the project engineer. Most of them were installed near the center line of the arch; others were placed in end or side walls where local conditions were monitored.

Relaxation Analysis of Power Plant Chamber

Any type of analysis of deformation around openings must be judged by how well or how poorly it represents the actual behavior of the rock.

One of the difficulties in a large opening is the collection of an adequate body of data to describe rock behavior. Restrictions on costs, personnel, and access during construction limit the number of measurements that can be made. Each measurement may represent 1000 ft^2 or more of excavated surface. There are obviously a number of other points within such an area that might have shown different deflections had they been measured. Nearly any body of deformation data obtained around openings includes some measurements which conform to the results expected from a given analysis and others which do not. Consistent correlations are rare, and it is often possible either to extol or malign a method of comparative analysis by applying data selectively. There are no methods of analyzing the behavior of rock masses for which this is not true, short of the millenium which envisions knowledge of the properties of all joints, fractures, blocks, and mineral compositions in the rock mass, together with the geometric descriptions needed for a computer analysis.

In presenting the following analysis of performance of the underground opening for Oroville power plant by the linear-elastic method previously described, all available measurements of rock movement are included in the analysis by representation of averages and ranges of deviation. It is shown that the performance was generally in accord with or superior to the performance predicted from rock properties measured during pre-construction testing. Reasons for superior performance in certain areas are considered. To the extent that an analysis can provide a basis for such considerations, it can be regarded as having been useful in evaluating the performance of the opening.

Deformations measured in the power plant chamber by borehole extensometers were evaluated by comparison with the relaxation analyses of the opening, which portray the absolute displacement projected to infinite depth. Extensometers of finite length necessarily record only a relative displacement that is a fraction of the total. In the case of the crown opening, the 20-ft extensometers could record about one quarter of the total displacement projected from the analysis. For the full opening, the 40-ft extensometers in the wall could record about one half of the total. These fractions assume that the deformation modulus throughout the rock mass is uniform. To the extent that the modulus varies from place to place, behavior would deviate from the analysis.

The relaxations measured at 20-ft depths during the excavation of the crown opening ranged from 0.002 to 0.059 in. and averaged 0.012 in. The average is compatible with the relaxation that would occur in a rock with a uniform modulus of 8.2×10^6 psi in the stress field shown in Fig. 6. The lowest modulus (1.6×10^6 psi) computed from the maximum relaxation is slightly larger than the conservative value adopted for design analyses. Figure 10 shows the depth-displacement curves for the power

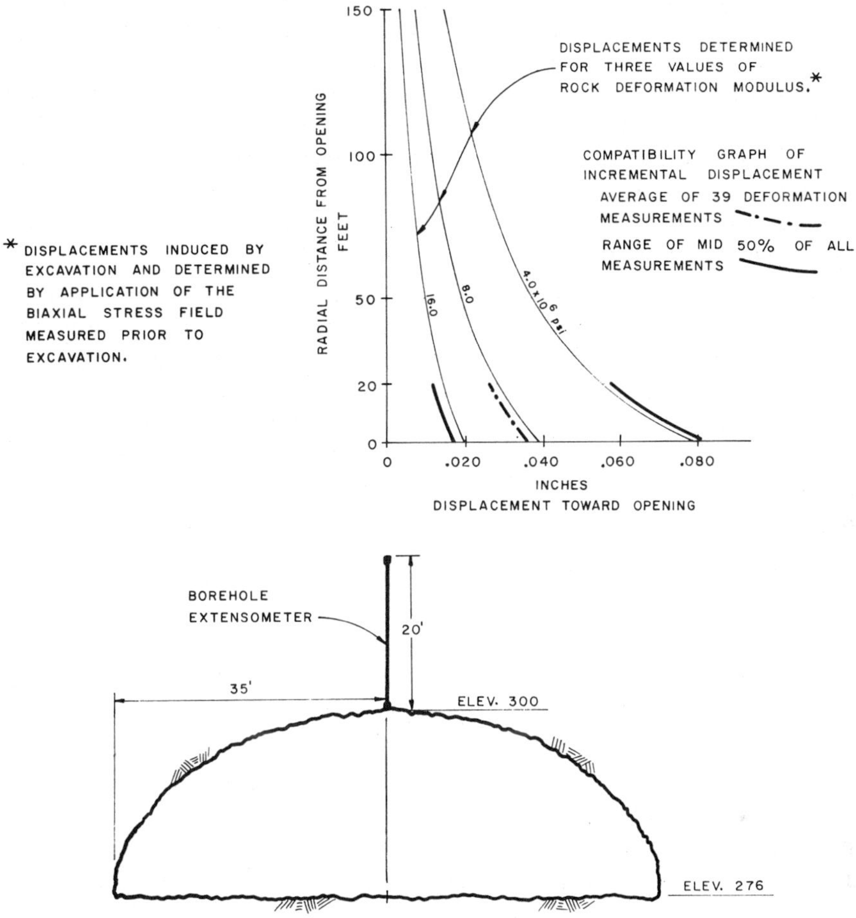

FIG. 10—*Correlation of arch deformation and rock modulus—Oroville underground power plant.*

plant arch projected for three values of rock modulus. Superimposed on these curves, in positions where their displacements and equivalent moduli are compatible, are the average relaxation measurement and the range of the middle 50 percent of the displacement recorded.

The small relaxations and equivalent high rock moduli encountered in the arch were remarkable. Before discussing this surprising performance, however, results obtained in the walls will be considered.

The walls of the chamber were reinforced with rock bolts on 6-ft centers. The contractor requested, and was granted, a revised bolting specification to permit benching in 12-ft lifts. The new specification extended the time for bolting from 3 to 48 h after each blast. Six 40-ft borehole extensometers with electrical strain gage sensors were installed near mid-height in each wall and at the center line of each of the six generating unit locations. At completion of the excavation, the relaxations recorded ranged from 0.054 to 0.247 in., and averaged 0.114 in. The average in both walls is compatible with the relaxation that would occur in rock of approximately 2×10^6-psi modulus. Depth-displacement graphs for each wall are shown in Fig. 11 for four values of rock modulus. The wall measurements are superimposed on these graphs in positions where the measured displacements and equivalent moduli are compatible.

Except for one measurement in the left wall, the performance at the walls conformed very closely to moduli measured in preliminary tests by the plate bearing and relaxation methods. The relatively large deflection in the left wall at the Generating Unit 2 center line was attributed to a geolological condition in which a prominent, nearly vertical shear intersected the wall at a shallow angle, striking across the borehole extensometer at shallow depth. The relatively large displacement is probably due to this shear. If this is true, the rock behavior is not linear over the entire length of the extensometer. Although the equivalent modulus for the 40-ft depth of rock is 0.8×10^6 psi, it is reasonable to assume that the deep rock has a higher modulus and the shallow rock a lower one. A consequence of this is that the 0.250-in. relaxation recorded by the extensometer is likely to represent more than half of the total displacement. Thus the total displacement is probably less than that projected on the graph. It is interesting to note that the strain gage sensors on the 40-ft extensometers were provided with a range of 0.5 in. in expansion, twice the maximum measurement encountered.

Direct wall-to-wall measurements would have been quite valuable in checking wall displacements, but such measurements were made only at the lower elevations in the opening, where access tunnels and auxiliary tunnels had already altered the rock mass prior to initial measurement. Subsequent wall-to-wall displacements measured at elevations 40 ft below the wall extensometers averaged 0.23 in.

Whereas the average modulus governing the crown deformations agreed closely with the average rock modulus determined from flatjack tests, the walls responded more nearly in accord with the lower moduli measured by the plate bearing and relaxation methods. Possible reasons for this have been suggested:

1. The stress field determined in preliminary testing and used throughout the analyses may be incorrect.

2. The rock may have anisotropic properties leading to vertical and horizontal disparities in behavior.

3. Borehole extensometers in the crown may have been installed too

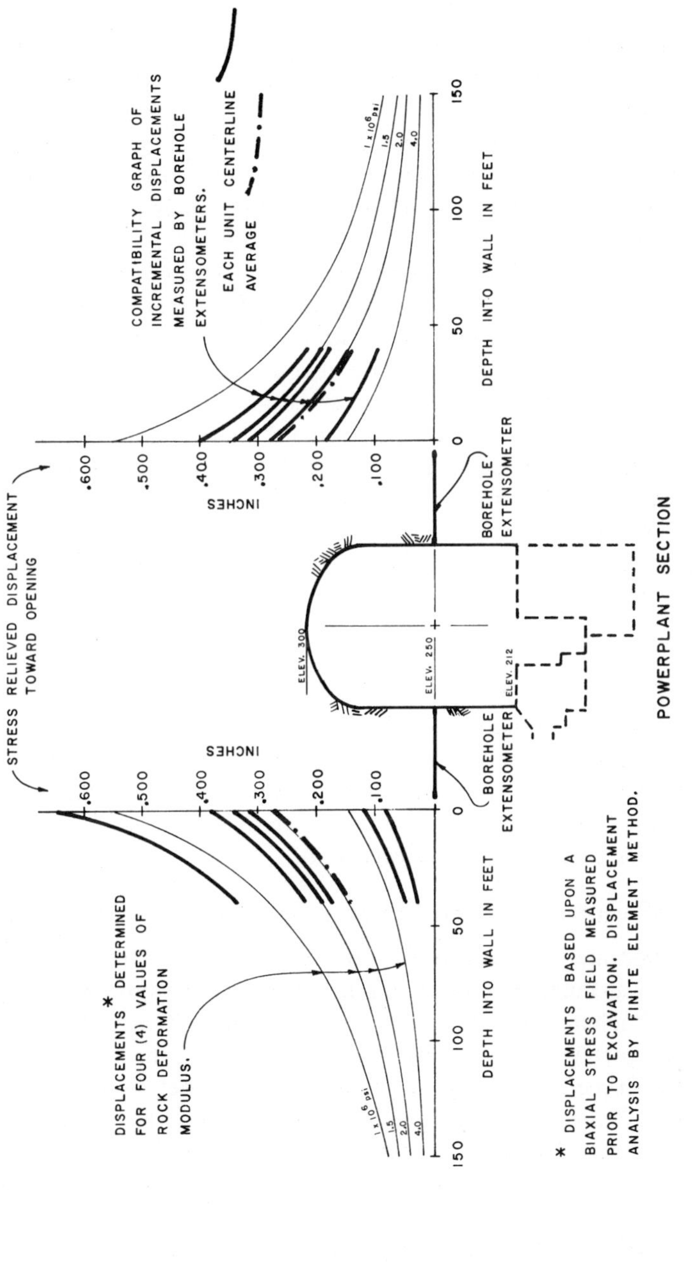

FIG. 11—*Correlation of wall deformation and rock modulus—Oroville underground power plant.*

far from the working headings and therefore measured appreciably less than the full relaxation.

4. An arch is a geometric shape with superior capabilities for resisting loads in compression.

5. The difference in bolt spacing and installation specifications may account for the difference in crown modulus and wall modulus.

The following comments are in response to the above suggestions:

1*a*. To account for the great difference in behavior between walls and arch by stress error, the lateral stress would have to be three or four times greater than the vertical. All available evidence firmly contradicts such a possibility.

2*a*. Some of the broken rock adjacent to shear zones exhibited anisotropic properties. However, the large-scale effect that would be required to explain the observed differences in wall and arch behavior was not evident. Neither did preconstruction testing reveal such behavior.

3*a*. The same construction personnel directed all extensometer installations, including those done earlier in the diversion tunnel. Records indicate that extensometers were installed from 1 to 3 ft of the working headings. Compared to the 70-ft span at the opening, this was very close.

4*a*. The relaxation analysis by the finite element method fully accounts for the geometry of the opening. The analysis shows that for a given rock modulus the crown would deflect about half as much as the wall. Arch deformations seven times smaller cannot be accounted for by the superior geometry of the arch.

5*a*. The superior bolt pattern and installation specification provides the most plausible explanation for the low arch deflections and equivalent high rock modulus. The modulus measured by flatjack tests (7.5×10^6 psi) was previously mentioned as a probable upper limit for the Oroville rock. If this upper limit was achieved in the arch, it is a singular tribute to the effectiveness of the rock reinforcement.

Rock Modulus Determination for Pressure Tunnel Design

Approximately a year after the excavation of Oroville power plant chamber was completed, a testing program to determine the rock modulus in the Tehachapi discharge lines was established. This feature of the State Water Project is a 1900-ft pumping lift in which the maximum designed pressures exceed 1100 psi. A profile of the tunnels is shown in Fig. 12. The tunnels are located in gneissic diorite rock, which is generally hard and strong but does contain zones of alteration. Three well-defined joint systems are present. The predominant joint set closely approximates the direction and dip of the foliation subparallel to the tunnel alinement and nearly vertical. The elastic modulus measured from cores varied from 1.3×10^6 to 15.6×10^6 psi, with an average of 5.8×10^6 psi.

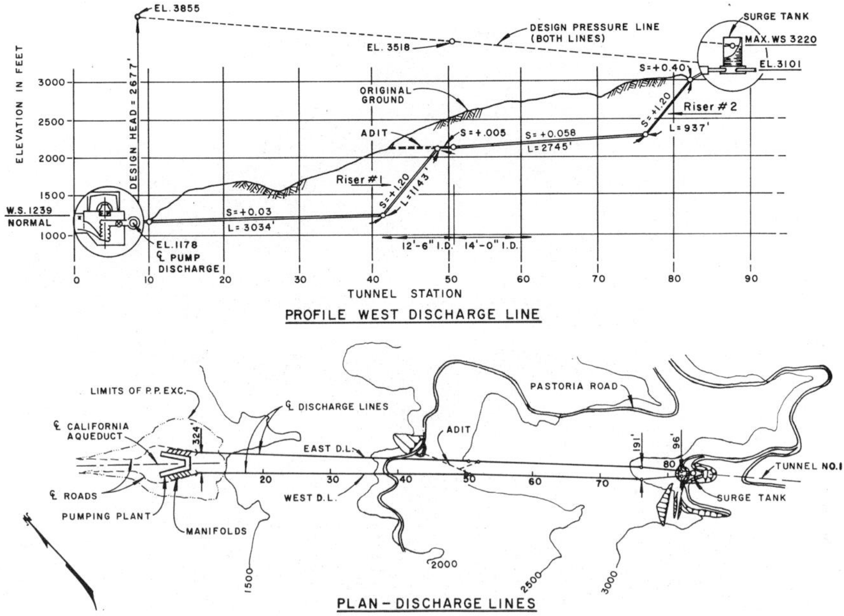

FIG. 12—*Plan and profile—Tehachapi discharge lines.*

Steel liners for the discharge lines were designed to act compositely with the backfill concrete and surrounding rock in such a manner that the rock support would prevent the liner from exceeding yield. Design procedures required that a deformation modulus for the rock be determined [*11*]. It was anticipated that the tunnel would penetrate zones of soft rock or sheared material that would require extra hoop reinforcement but that the steel liner thickness would be determined by the average rock modulus. It was necessary because of scheduling considerations to establish plate thickness in advance of tunnel excavation. Testing procedures employed two dissimilar methods to measure the rock modulus. Conventional plate bearing tests were chosen for a primary method because of their demonstrated record of successful use.

A 500-ton-capacity jacking assembly fabricated from high strength aluminum was designed for the Tehachapi tests. The jacks were similar to those used for uniaxial tests by the Bureau of Reclamation [*12*], but with some modifications. A rigid plate loading was used in which the load was applied by a single flatjack near midspan of the jacking assembly. A photograph of the assembled equipment in the test adit is shown in Fig. 13. Bearing plates of high strength aluminum were 3 in. thick and 21 in. in diameter. Average pressure at the rock was approximately 2500 psi maximum. Thirty-two measurements were made at 16 sites. A typical load-deformation curve is shown in Fig. 14.

FIG. 13—*500-ton jacking equipment for plate bearing tests.*

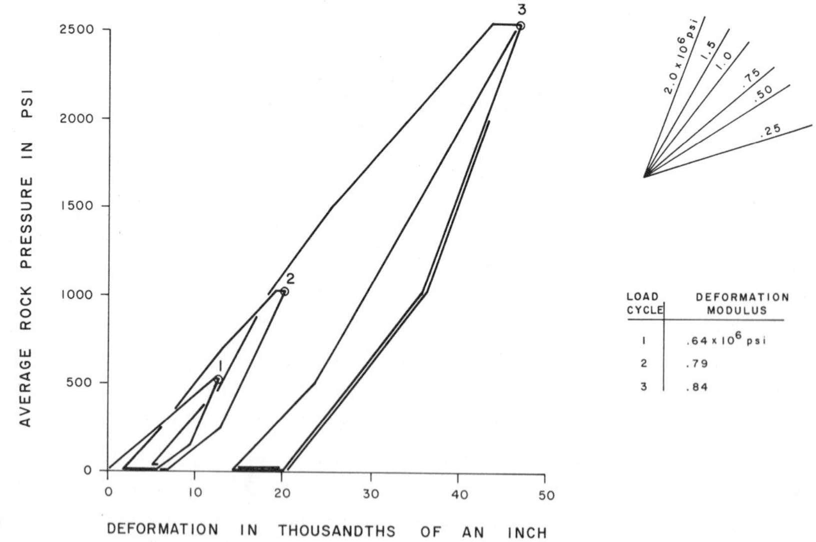

FIG. 14—*Typical load-deformation curves, plate bearing tests—Tehachapi discharge lines.*

A secondary method of measuring the rock modulus, for corroboration of the plate bearing tests, was the subject of considerable study. Pressure chamber tests were considered, along with the possibility of utilizing the relaxation analysis which had been used at Oroville power plant. Relaxation tests were chosen in lieu of pressure chamber tests because of the greater flexibility in evaluating the rock at frequent intervals along the entire alinement of the tunnels. Relaxation tests also afford a more rapid method of obtaining the rock modulus.

Improved borehole extensometers featuring twin-span measurements for intervals of 0 to 5 ft and 5 to 15 ft were designed for the Tehachapi tests. Measuring rods attached to two deep anchors were directed into a stainless steel measuring head as shown in Fig. 15. A tape extensometer

FIG. 15—*Twin-span borehole extensometer measuring head.*

(Fig. 16) was developed for wall-to-wall measurements using the extensometer heads as references. These devices permitted the continuous measurement of tunnel relaxation over a span of 46 ft, including the 16-ft tunnel diameter. It was expected that this technique would monitor a larger proportion of the total displacement than was measured at Oroville.

Stress measurements by the borehole relief method were made in the upper and lower reaches of the tunnels. The amount of cover over the tunnels does not vary greatly except between Stations 35 to 40. Stress determinations were generally in agreement with the overburden, with

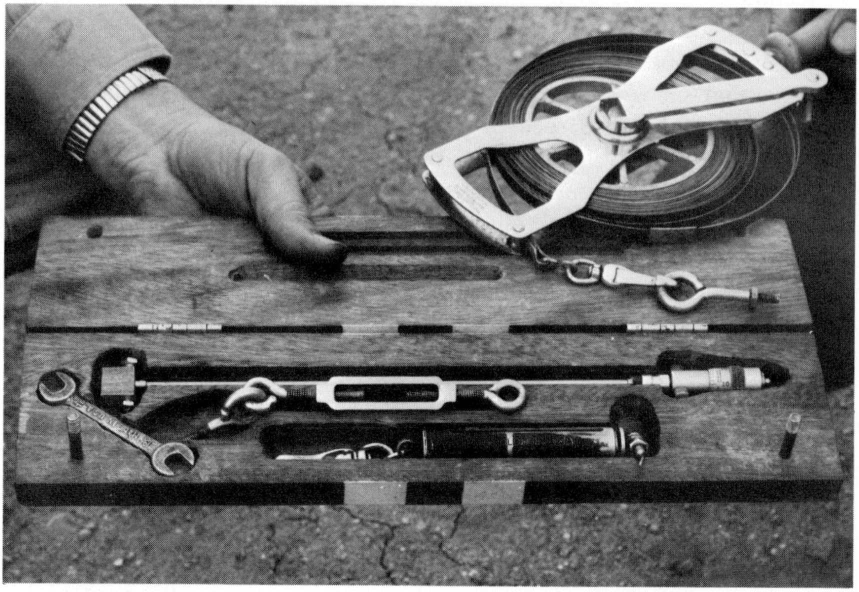

FIG. 16—*Tape extensometer.*

lateral stresses equal to or slightly less than overburden. Stress measurements were more difficult to make in the diorite rock than in the Oroville amphibolite. For the ten stress relief holes attempted, it was possible to obtain continuous records in only five. Typical measurements are shown in Fig. 17.

It was not practical to install the twin-span borehole extensometers from the drilling jumbo because of the greater time required for their installation. Therefore, the earlier procedure in which a ring of instruments was installed by Department personnel following the last work shift of the week was adopted. This precluded any possibility of selecting sites; but, when the heading was in soft or sheared rock, installations were omitted. Occasionally only one or two extensometers were installed at a station. Approximately 125 extensometers were installed at 50 stations in the tunnels.

Extensometer readings were widely scattered and occasional negative readings were encountered, indicating compressive displacement. As pointed out earlier, rock blocks may move in translation or rotation; thus, where individual measurements often prove to be anomalous, averages of an increasingly large sample soon define a consistent result which does not appreciably change as more measurements are added.

Typical time-deformation curves plotted from extensometer measurements and tunnel span measurements are shown in Fig. 18. The corres-

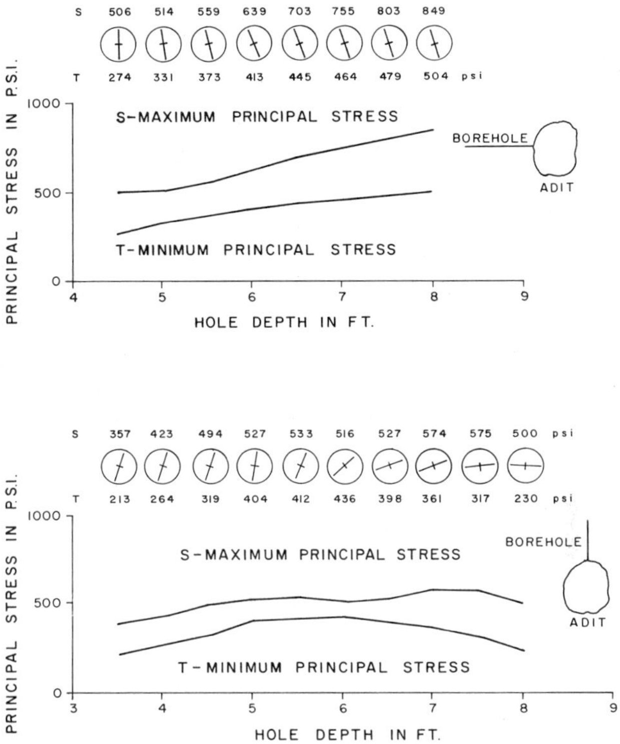

FIG. 17—*Typical borehole stress relief measurements—Tehachapi adit.*

ponding relaxation analysis is shown in Fig. 19. It can be seen that the algebraic sum of all the displacements across the horizontal 46-ft span leaves a residual amount of displacement beyond the reach of the extensometers unmeasured. Since no measurement of the tunnel span was practical in the vertical direction, crown relaxations were analyzed in the manner used for the Oroville relaxations. The modulus in the crown, computed from the 15-ft relaxation, is shown in Fig. 19 in the position where the modulus is compatible with the relaxation measured.

A summary of the deformation moduli determined from the relaxations and by plate bearing tests is portrayed in Fig. 20. The peaks of the curves represent the median values for the respective tests, while the breadth of the curves represents the scatter of results. For example, the median determination of the rock modulus from plate bearing tests is 0.5×10^6 psi, and the mid 50 percent of all such results ranges from approximately 0.35×10^6 to 0.7×10^6 psi. The median result from relaxation tests is 0.6×10^6 psi, and the mid 50 percent of the tests ranges from approximately 0.3×10^6 to 1.8×10^6 psi.

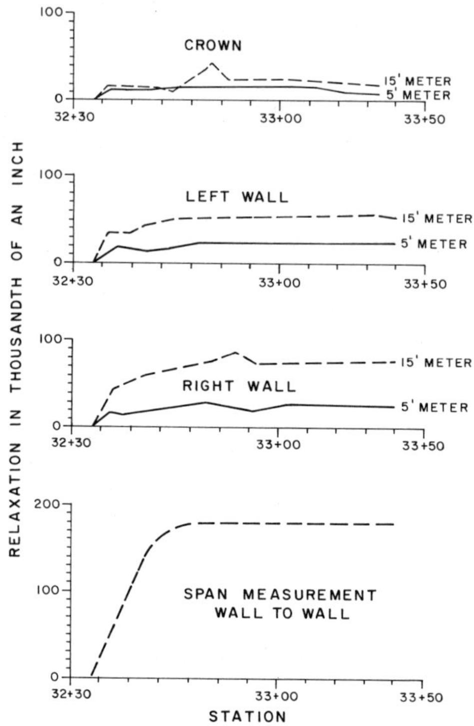

FIG. 18—*Deformation versus tunnel heading advance—Station 32 + 33 west, Tehachapi discharge lines.*

An additional comparison is shown in Fig. 21, which contrasts the results obtained by relaxation analyses of the walls with that of the crown. Where the median determination of modulus for the walls was slightly under 0.5×10^6 psi, the crown relaxations produced a median result of 1.5×10^6 psi. Unlike the results in Oroville power plant, however, this condition is attributed to the anisotropic character of rock, which is due to the predominant joint set mentioned earlier. A more precise analysis of this condition could be obtained by developing a nonlinear relaxation analysis embracing the anisotropy of the joints.

Still another comparison of the deformation moduli determined from relaxation at depths of 0 to 5 ft and 5 to 15 ft. The results, shown in Fig. 22, indicate that there is little improvement of the modulus with depth.

Summary and Conclusions

A method of analysis was presented by which the relaxations which accompany the driving of a tunnel in competent rock are used to determine the deformation modulus when the rock stresses are known. The

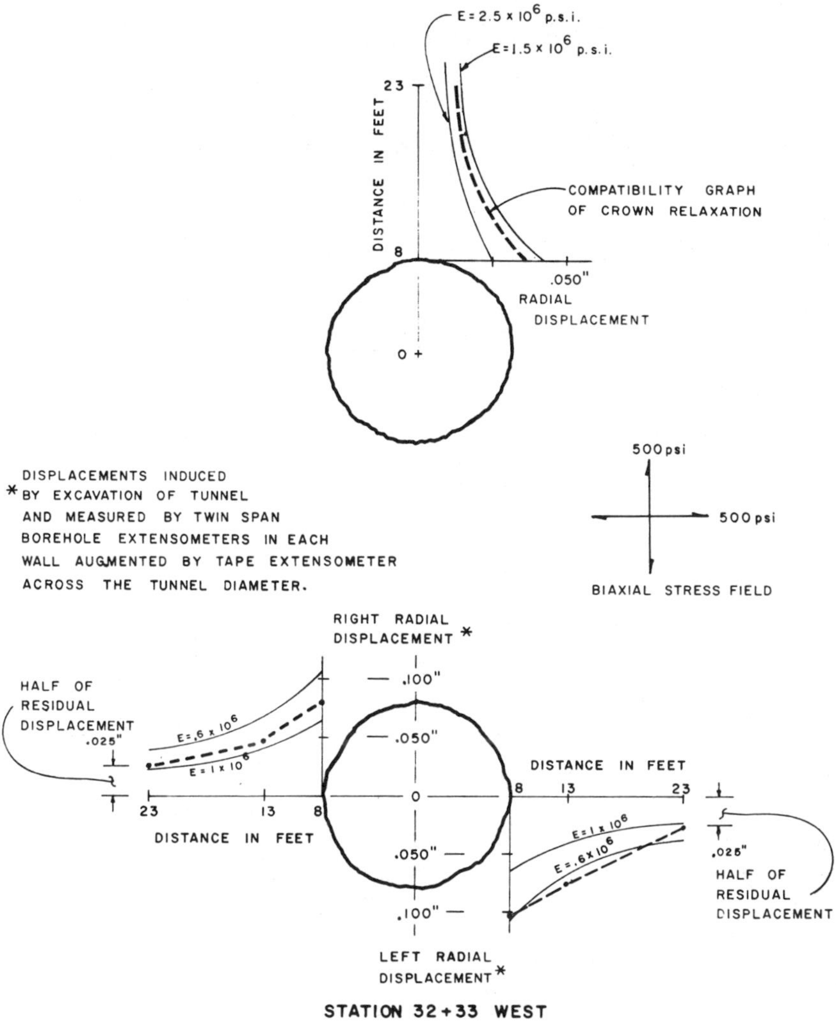

FIG. 19—*Correlation of relaxation and rock modulus—Tehachapi discharge lines.*

method has been used as a supplementary analysis for evaluating rock behavior in an underground power plant and a steel-lined pressure tunnel. All methods presently used for *in situ* determination of rock modulus require that a load be applied to the rock and the resulting displacement measured. The method of relaxation analysis presented in this paper enlarges the scope of rock modulus determinations by using rock displacements induced by load removal.

The magnitude of load removal accomplished in the excavation of a tunnel dwarfs any loads presently used in modulus testing, even by the

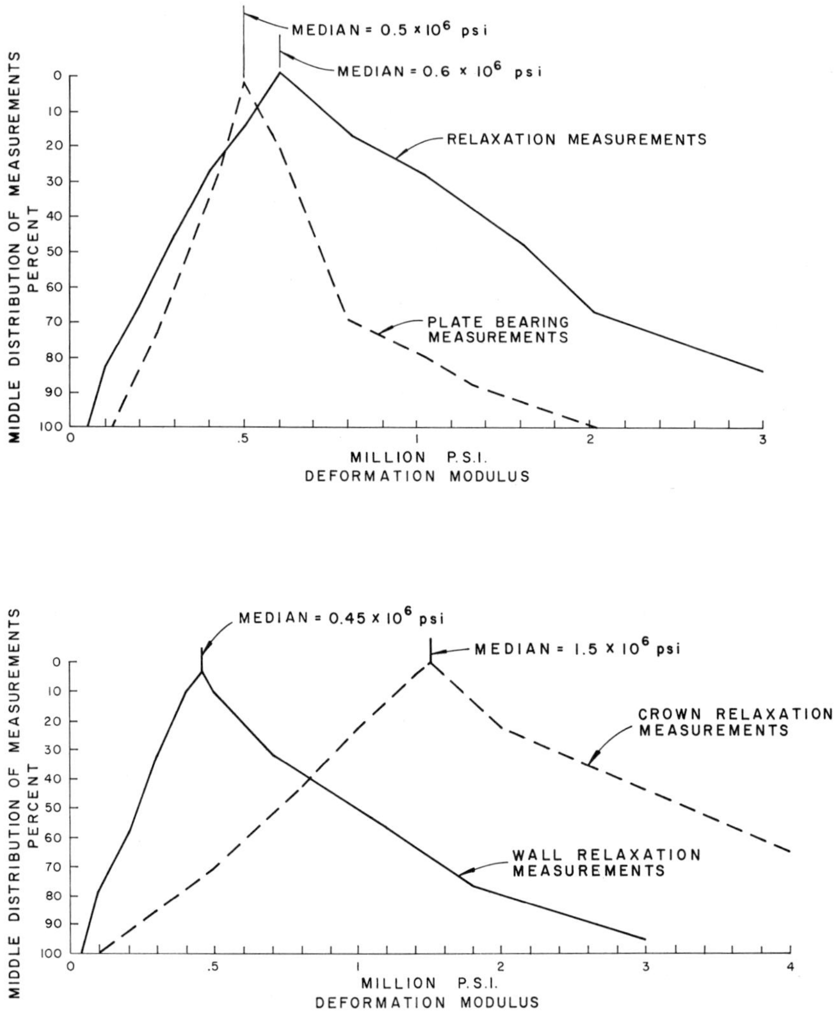

FIG. 21—*Summary, rock deformation modulus measurements—Tehachapi discharge lines.*

largest of test procedures. Loads measured in the millions of tons are released, and rock volumes measured in the hundreds of thousands of cubic yards are subjected to minute but measurable deformations. With such a fertile source of information about rock behavior at hand, it is inevitable that new techniques should evolve to exploit this massive phenomenon.

Although the greatest advantage of the method lies in the unprecedented scale of measurement, there are other advantages. Where other methods

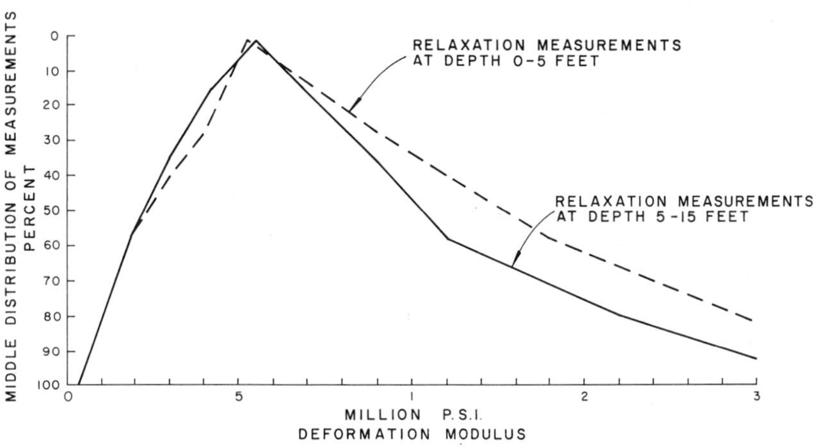

FIG. 22—*Summary, rock deformation modulus measurements—Tehachapi discharge lines.*

of testing begin with the second cycle of deformation, relaxations around openings represent the first cycle. The method requires a minimum investment in equipment and permits a rapid determination of the modulus in large tunnels and chambers. The effectiveness of the method requires a large sampling of relaxation data and, therefore, places emphasis on low-cost borehole extensometers which can easily and quickly be installed, preferably from a drilling jumbo. Extensometers should be relatively immune to blast damage but need not have unusually great measurement precision. Diametral span measurements contiguous with extensometer measurements facilitate analyses and are desirable. Multiple extensometer positions augment the data and are desirable if installation procedures are kept reasonably simple.

The determinations of primary stress needed for the evaluation of the rock modulus from tunnel relaxations would not have been possible prior to the development of borehole stress relief techniques and equipment. Further improvements in these techniques have occurred during the time these projects were under construction, and their availability will further enhance the feasibility of relaxation analyses.

APPENDIX

u = radial displacement resulting from application of biaxial stress to an infinite plate with a circular hole.[7]

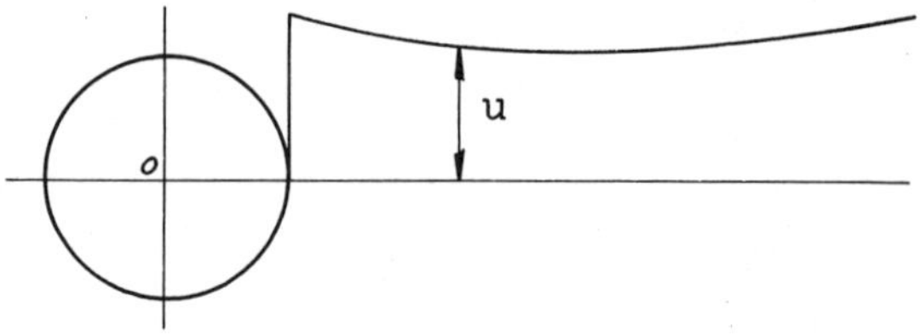

u_o = radial displacement resulting from application of biaxial stress to a continuous and solid plate.

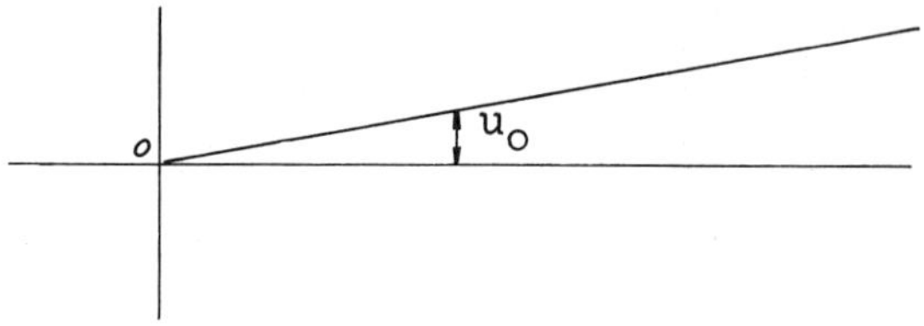

$u_x = u - u_o$

u_x = radial displacement resulting from forming a circular hole in an already stressed plate.

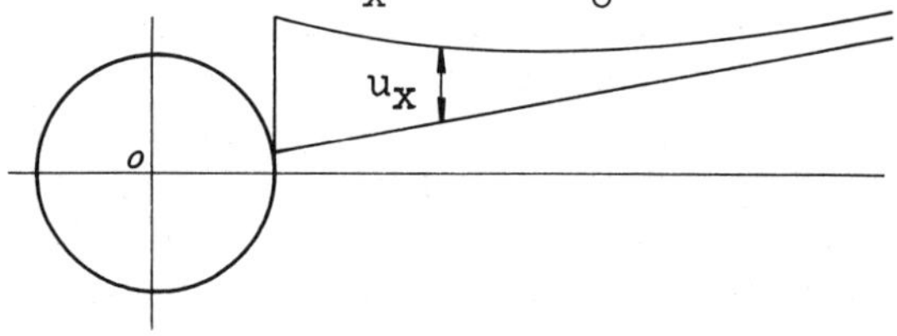

a = radius of hole

r, θ = polar parameters

E = elastic modulus

S, Q = perpendicularly applied stresses

ν = Poisson's ratio

[7] The derivation for the displacement u has been published in "Deformation of a Borehole in Rock," R. H. Merrill and J. R. Petersen, RI 5881, U.S. Bureau of Mines, 1961, p. 26.

The equation modified for plane strain and biaxial stress is:

$$u = \frac{1-\nu^2}{E}\left[\frac{S+Q}{2}\left(r+\frac{a^2}{r}\right)+\frac{S-Q}{2}\left(r+\frac{4a^2}{r}-\frac{a^4}{r^3}\right)\cos 2\theta\right]$$

$$-\frac{\nu(1+\nu)}{E}\left[\frac{S+Q}{2}\left(r-\frac{a^2}{r}\right)-\frac{S-Q}{2}\left(r-\frac{a^4}{r^3}\right)\cos 2\theta\right] \quad \text{.............(1)}$$

The displacement u_0 can be found from Eq 1 by setting $a = 0$:

$$u_0 = \frac{1-\nu^2}{E}\left[\frac{S+Q}{2}(r)+\frac{S-Q}{2}(r\cos 2\theta)\right]$$

$$-\frac{\nu(1+\nu)}{E}\left[\frac{S+Q}{2}(r)-\frac{S-Q}{2}(r\cos 2\theta)\right] \quad \text{.............(2)}$$

Since $u_x = u - u_0$,

$$u_x = \frac{1-\nu^2}{E}\left[\frac{S+Q}{2}\left(\frac{a^2}{r}\right)+\frac{S-Q}{2}\left(\frac{4a^2}{r}-\frac{a^4}{r^3}\right)\cos 2\theta\right]$$

$$+\frac{\nu(1+\nu)}{E}\left[\frac{S+Q}{2}\left(\frac{a^2}{r}\right)-\frac{S-Q}{2}\left(\frac{a^4}{r^3}\right)\cos 2\theta\right]$$

and simplifying,

$$u_x = \frac{Sa^2}{2Er}\left[(1+\nu)\left(1-\frac{a^2}{r^2}\cos 2\theta\right)+4(1-\nu^2)\cos 2\theta\right]$$

$$+\frac{Qa^2}{2Er}(1+\nu)\left(1+\frac{a^2}{r^2}\cos 2\theta\right)-4(1-\nu^2)\cos 2\theta\Bigg] \quad \text{.............(3)}$$

A similar analysis for an elliptical hole is based upon Eq 99 published in *Theory of Elasticity,* Timoshenko and Goodier, McGraw-Hill, 1961:

$$2G(u+iv) = \frac{3-\nu}{1+\nu}\psi(z) - z\overline{\psi}'(\bar{z}) - \bar{\chi}'(\bar{z}) \quad \text{.............(4)}$$

Graphs for "relaxation" displacements in various stress fields are shown for both circular and elliptical holes in Figs. 1 through 4.

References

[1] Waldorf, W. A., Veltrop, J. A. and Curtiss, J. J., Paper 3576, *Proceedings,* American Society of Civil Engineers, PACEA, SM4, July 1963.

[2] Obert, L. and Duvall, W. J., *Rock Mechanics and the Design of Structures in Rock,* John Wiley & Sons, New York, 1967.

[*3*] Merrill, R. H. and Peterson, J. R., "Deformation of a Borehole in Rock," U. S. Bureau of Mines Report of Investigations 5881.

[*4*] Alexander, I. G., "Field and Laboratory Tests in Rock Mechanics," Third Australia-New Zealand Conference on Soil Mechanics and Foundation Engineering, 1960.

[*5*] Schilling, A. A., Paper 5150, *Proceedings,* American Society of Civil Engineering, PACEA, Vol. 93, PO1, March 1967.

[*6*] Grosvenor, N. E. and Abel, J. F., "Measurement on the Pilot Bore for the Straight Creek Tunnel," Highway Research Board Publication No. 1379, 1966.

[*7*] Seery, J. D., Paper 5154, *Proceedings,* American Society of Civil Engineering, PACEA, Vol. 93, PO1, March 1967.

[*8*] Merrill, R. H., Williamson, J. V., Ropchan, D. M. and Kruse, G. H., "Stress Determinations by Flatjack and Borehole Deformation Methods," U. S. Bureau of Mines Report of Investigations 6400, 1964.

[*9*] Thayer, D. P., Stroppini, E. W. and Kruse, G. H., "Properties of Rock at Underground Power Plant, Oroville Dam," Eighth International Conference on Large Dams, Edinburgh, Scotland, 1964.

[*10*] "Plate Bearing Tests at Oroville Powerhouse Site," Rock Mechanics Report RM-3, California Department of Water Resources, 1963.

[*11*] Vaughn, E. W., Power Division Paper 949, *Proceedings,* American Society of Civil Engineers, PACEA, April 1956.

[*12*] Wallace, G. B., Slebir, E. J. and Anderson, F. A., "Radial Jacking Test for Arch Dams," Tenth Rock Mechanics Symposium, University of Texas, Austin, Tex., May 1968.

R. P. Benson,[1] *D. K. Murphy,*[1] *and D. R. McCreath*[1]

Modulus Testing of Rock at the Churchill Falls Underground Powerhouse, Labrador

REFERENCE: Benson, R. P., Murphy, D. K. and McCreath, D. R., **"Modulus Testing of Rock at the Churchill Falls Underground Powerhouse, Labrador,"** *Determination of the* In Situ *Modulus of Deformation of Rock, ASTM STP 477,* American Society for Testing and Materials, 1970, pp. 89–116.

ABSTRACT: At Churchill Falls, the rocks comprise a relatively homogeneous, high-quality gneissic assemblage. Modulus testing was performed on intact laboratory specimens and on *in situ* mass rock at a depth of 1000 ft; standard equations of elasticity were used to calculate the moduli from the tests, although deviance from elastic behavior was noted, especially with the *in situ* plate jack tests. Plate jack tests were strongly affected by the compressibility of blast-damaged rock around the underground openings, which resulted in inelastic behavior and a low modulus for rock near the surface of the opening, but a high modulus and approximately elastic behavior for rock at depth. These effects are particularly important in the design of pressure conduits.

The results of modulus tests reflect the effects of structural discontinuities in, and the size of, the specimen, as well as its *in situ* and applied stress environment. Accordingly, the testing apparatus and methods must be carefully selected to accurately measure rock behavior relative to the expected mode of behavior in design.

KEY WORDS: tests, laboratories, instruments, modulus of elasticity, deformation, stresses, rocks, boreholes, subsurface structures

With the growing awareness of rock-structure interaction problems, testing is now being used to assess the deformation response of mass rock subjected to known loads. In general, testing has been related to the design of foundations for concrete structures such as dams and heavy

[1] Specialist consultant, Rock Mechanics, specialist consultant, Engineering Geology, and senior engineer, Rock Mechanics, respectively, H. G. Acres Ltd., Niagara Falls, Ontario, Canada.

buildings; but more recently it has been used to assess the behavior of rock surrounding underground structures such as pressure conduits, powerhouses, and mines.

Exploration relating to design of the underground power chambers at Churchill Falls, Labrador, has included laboratory modulus tests on intact specimens and plate jack tests to determine the modulus of rock *in situ*. The purpose of the testing was twofold: (1) to obtain data on the behavior of mass rock for use in certain aspects of design, such as the steel-lined penstocks, and (2) to develop a model, based on finite element stress analysis, to which the deformation response of the major openings can be related as excavation proceeds.

Site Geology

The project is located on the Central Labrador Plateau, an area of generally low relief underlain by Precambrian metamorphic and igneous rocks presumed to be of Grenville age. The area, shown in Fig. 1, is

FIG. 1—*Location map.*

covered with a thin mantle of glacial till and outwash sediments derived from the most recent glaciation of the Pleistocene epoch.

The most striking physical feature in the area is the Churchill River Valley, a preglacial valley approximately 200 miles long, scoured and deepened by glaciation to a depth of 800 to 1000 ft below the general level of the plateau. In the immediate area of the underground power installations, the rocks comprise a gneissic assemblage intruded by diorite-gabbro and lesser amounts of granite, syenite, and pegmatite. All of the rocks are hard and crystalline with poorly developed gneissic foliation that trends east to west, parallel with the river and subparallel with the main underground openings. Foliation has a consistent and steep southerly dip; although on a very local scale, considerable variation is observed.

The major structural features in the area of the proposed power installations were identified by air photo interpretation, surface mapping, and exploratory drilling. Rock quality studies performed during the exploration phase of the project included:

1. Geological interpretation of rock cores.
2. Statistical and qualitative joint surveys of surface outcrops and borehole camera studies in deep holes.
3. Water pressure testing in boreholes.
4. Laboratory testing of core specimens.

The rock in the area of the underground complex has four predominant joint sets. The joints are tight, and, where alteration products are present, consist mainly of very thin coatings of hematite or gypsum. Other alteration products or joint-filling materials identified include laumontite, chlorite, calcite, and metabentonite. A notable feature, first indicated in the borehole photographs and later verified in the excavations, is the preferential coating of the near-horizontal joint set with gypsum, which most commonly occurs as a discontinuous white film.

Laboratory testing showed the major rock types to be similar in engineering characteristics with an average uniaxial compressive strength of 16,000 psi, as shown in Fig. 2. An exception is the foliated biotite gneiss, which indicated strengths as low as 6000 psi due to the weak micaceous layers in the specimens. However, it was expected that this material would only be encountered locally in the excavations.

All studies indicated that the powerhouse would be excavated in hard and durable rock of high quality and generally free of major zones of weakness. With the relatively homogeneous rock types and their similarity in behavior and quality, it was further considered that the rock mass would respond in an elastic fashion, as has been experienced at Morrow Point [*1*],[2] Tumut [*2*], and the Nevada Test Site [*3*]. In this regard, the

[2] Italic numbers in brackets refer to the list of references at the end of this paper.

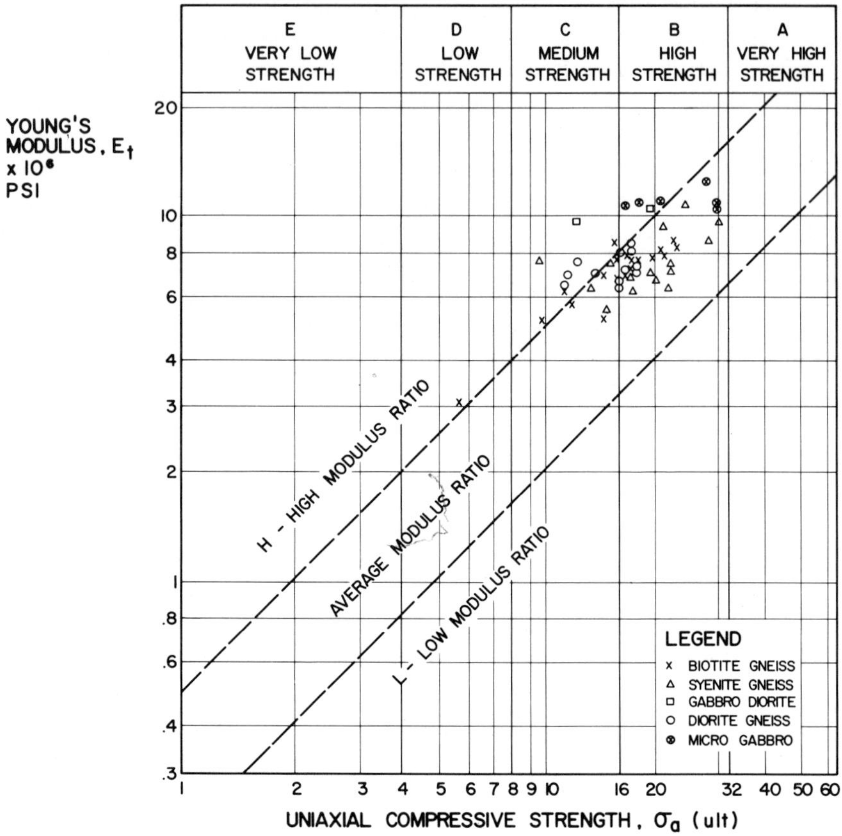

FIG. 2—*Engineering classification for intact rock (Churchill Falls).*

elastic finite element approach has been extensively used as an aid to design. The mathematical techniques of this method have been outlined in a previous publication, Sigvaldason et al [*4*].

Plate Jack Testing

Location of Tests

Provision for test drifts (tunnels) was incorporated in the initial excavation contract which provided first access to the underground area of the main complex. The locations of test drifts were selected to be representative of the rock conditions and to be as close as possible to the powerhouse. The drifts were approximately 7 ft by 8 ft by 80 ft long, and their locations relative to major geological features are shown on Fig. 3.

Drifts 2 and 3 were excavated in rock considered to be representative of the major portion of rock encountered in the excavations. These drifts

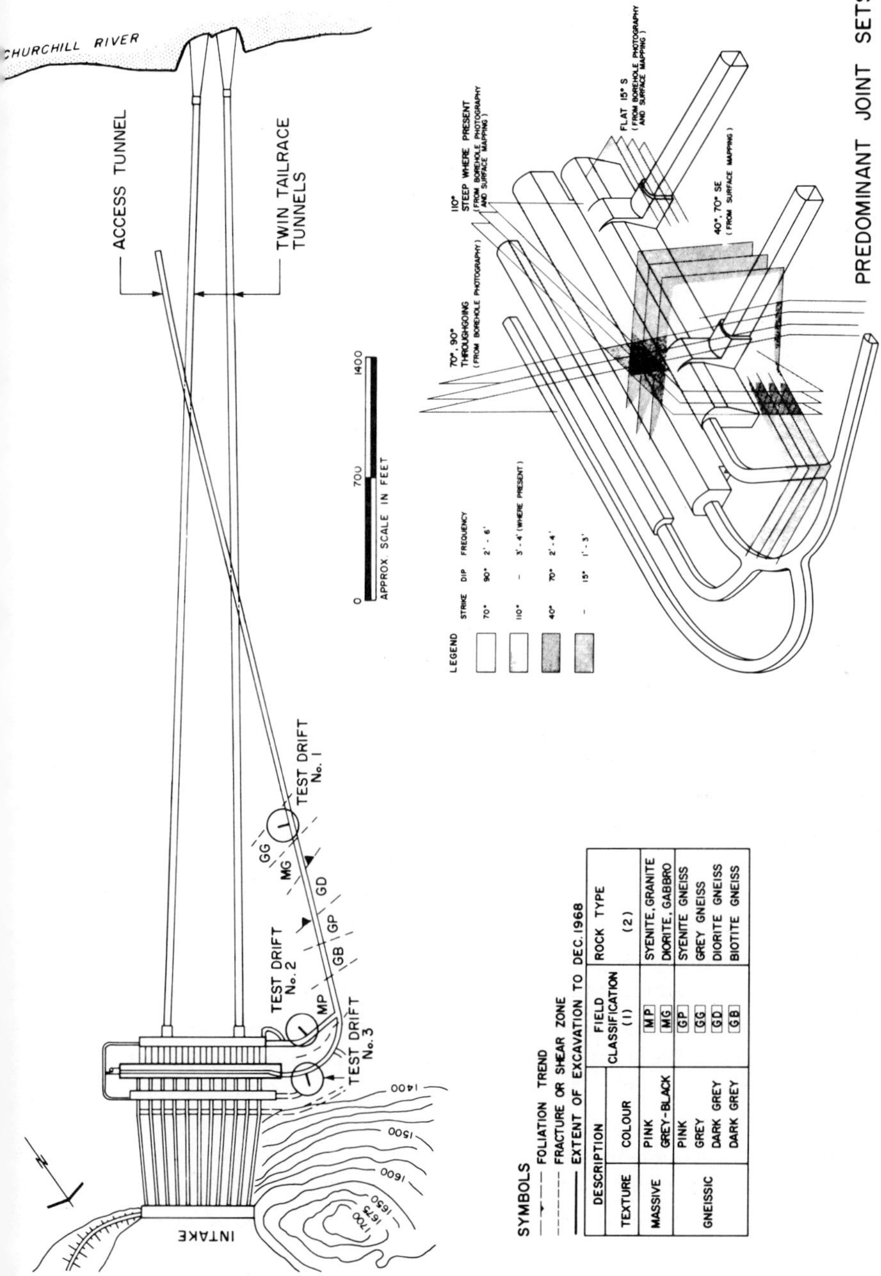

DESCRIPTION		FIELD CLASSIFICATION	ROCK TYPE
TEXTURE	COLOUR	(1)	(2)
MASSIVE	PINK	MP	SYENITE, GRANITE
	GREY-BLACK	MG	DIORITE, GABBRO
GNEISSIC	PINK	GP	SYENITE GNEISS
	GREY	GG	GREY GNEISS
	DARK GREY	GD	DIORITE GNEISS
	DARK GREY	GB	BIOTITE GNEISS

FIG. 3—*Powerhouse complex. Geology and location of test drifts.*

are approximately normal to the foliation, which is moderate to faint. The rock is generally massive, with widely spaced, tight joints. Drift 1 was driven in an area of minor shearing, representative of a lower quality rock that will be encountered locally in the excavations.

Rock Preparation

The drifts were excavated by controlled perimeter blasting using a 12-in. hole spacing and low density powder to minimize rock damage to the walls. The precise locations for the jacks were then marked out at least 50 ft from the drift entrance and where rock conditions were compatible with average rock quality. Excavation of the load areas was done with percussion drills and chipping hammers. Each load area was perpendicular to the intended axis of the jacking unit, and local irregularities were limited to 2 in. Up to 2 ft of rock was excavated to remove blast-damaged material, although minute cracks parallel to the face could still be seen in the drill holes within the first 6 to 12 in. of rock behind the plate jacks.

Typical Test Setup

BX holes were drilled to a depth of 16 ft behind each flatjack. The core was then examined and logged to determine the best location for each anchor relative to joints or cracks in the core. Results of the logging are shown in Figs. 4 to 6. Both electrical and mechanical extensometers were used in these holes to determine the rock deformation. Several anchor points were used in each hole (Fig. 7) so that the expected variation in the modulus with depth could be calculated.

Electrical extensometers with SR4 gages mounted on moveable cantilevers were used to measure the deformations behind the center of the jack. Each cantilever was attached by piano wire to a rock anchor, with a maximum of eight anchors per extensometer. These were complemented with similar eight-position mechanical extensometers placed at the edge of the jack. Mechanical extensometers were also used to determine the movement between the plate jacks and the relative movement of each wall of the drift. Initial tests showed that the mechanical extensometers placed at the edge of the jack were not sensitive enough to accurately measure rock movement because very small deformations were induced in the area. Their use was discontinued, and in some cases electrical extensometers were substituted to obtain more precise measurements.

The jack equipment was then placed into position. The basic installation included four segmented aluminum columns to span the opening, aluminum heads bearing against flexible, oil-filled flatjacks of 34-in. diameter, and mortar pads to carry the thrust from the jacks to the rock. Refinements included transverse stiffeners for the columns, screw-type feet at the

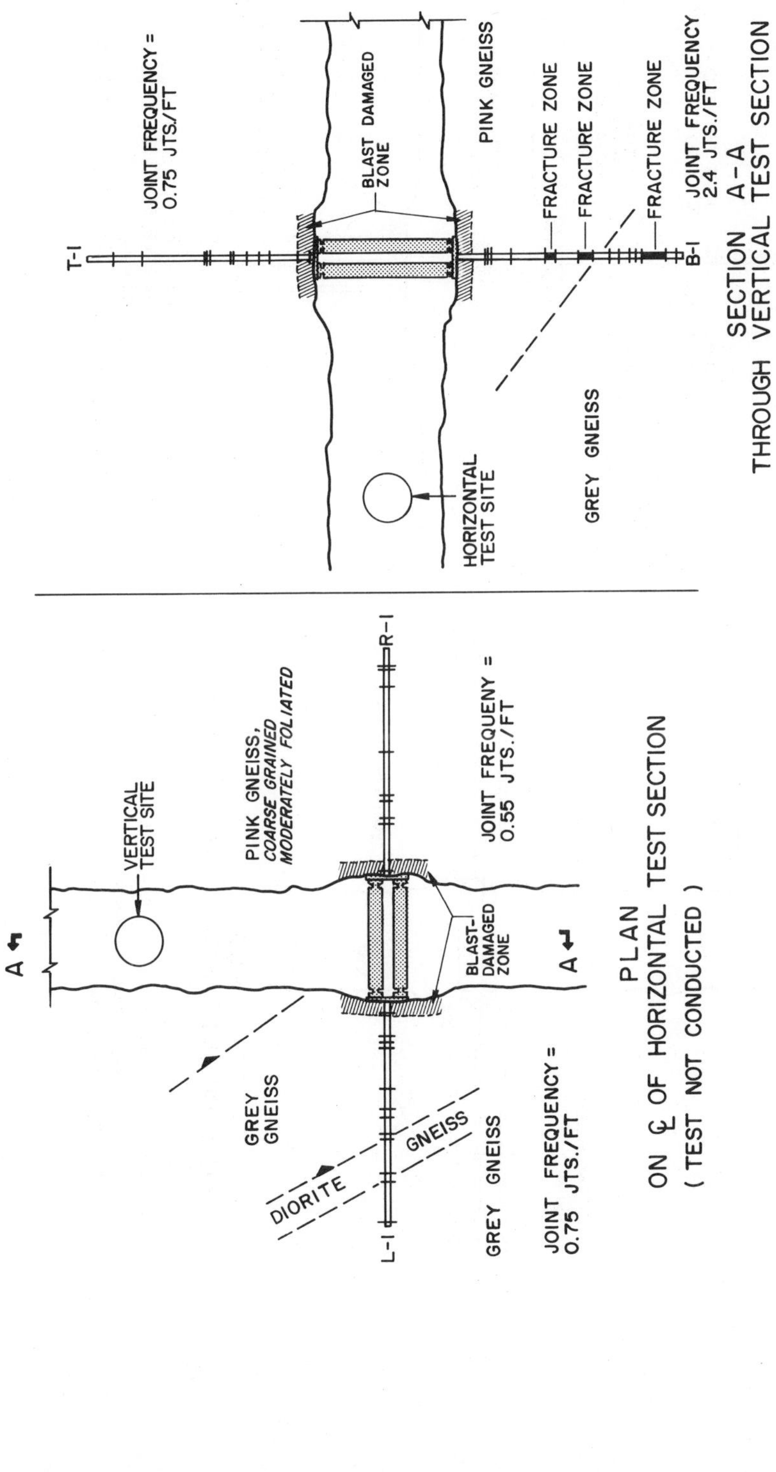

FIG. 4—*Test Drift 1 jacking sites—rock quality.*

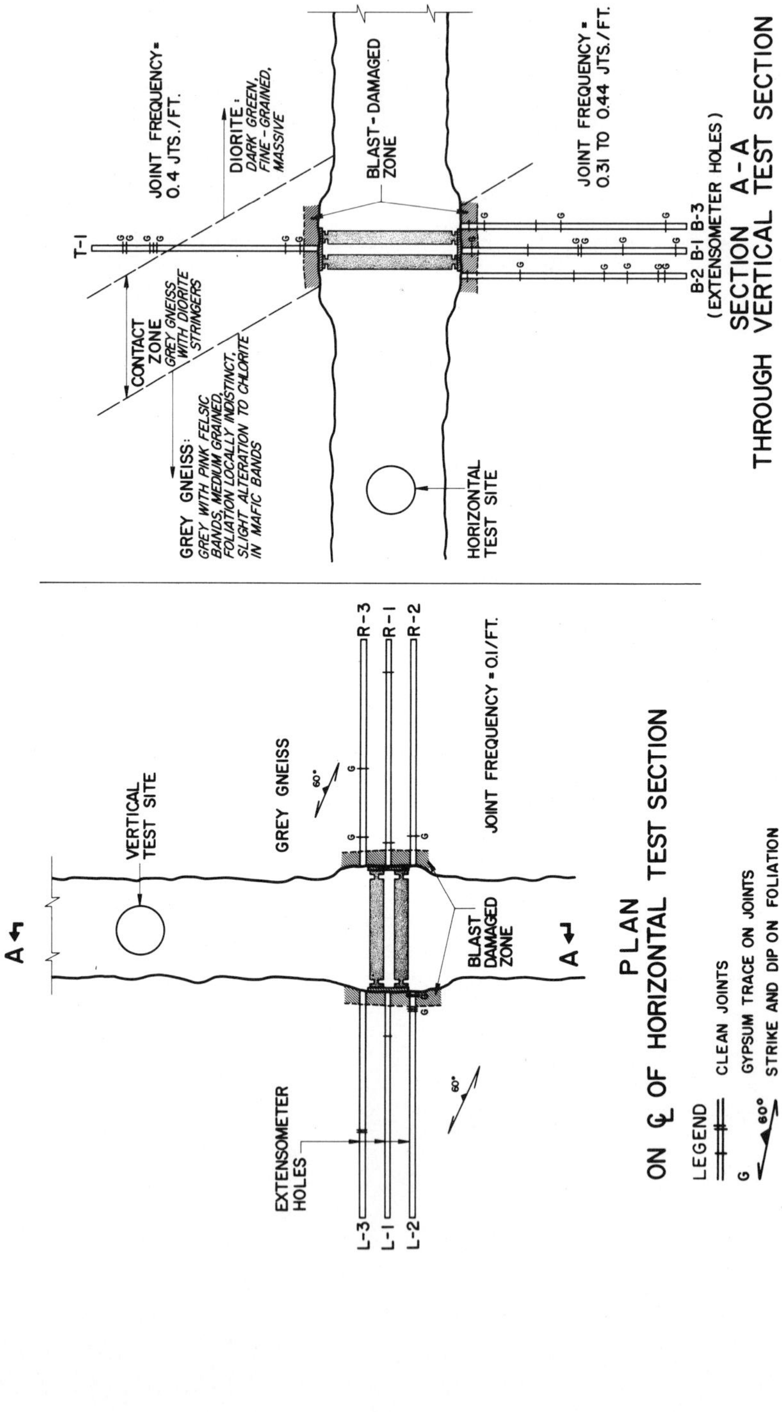

FIG. 5—*Test Drift 2 jacking sites—rock quality.*

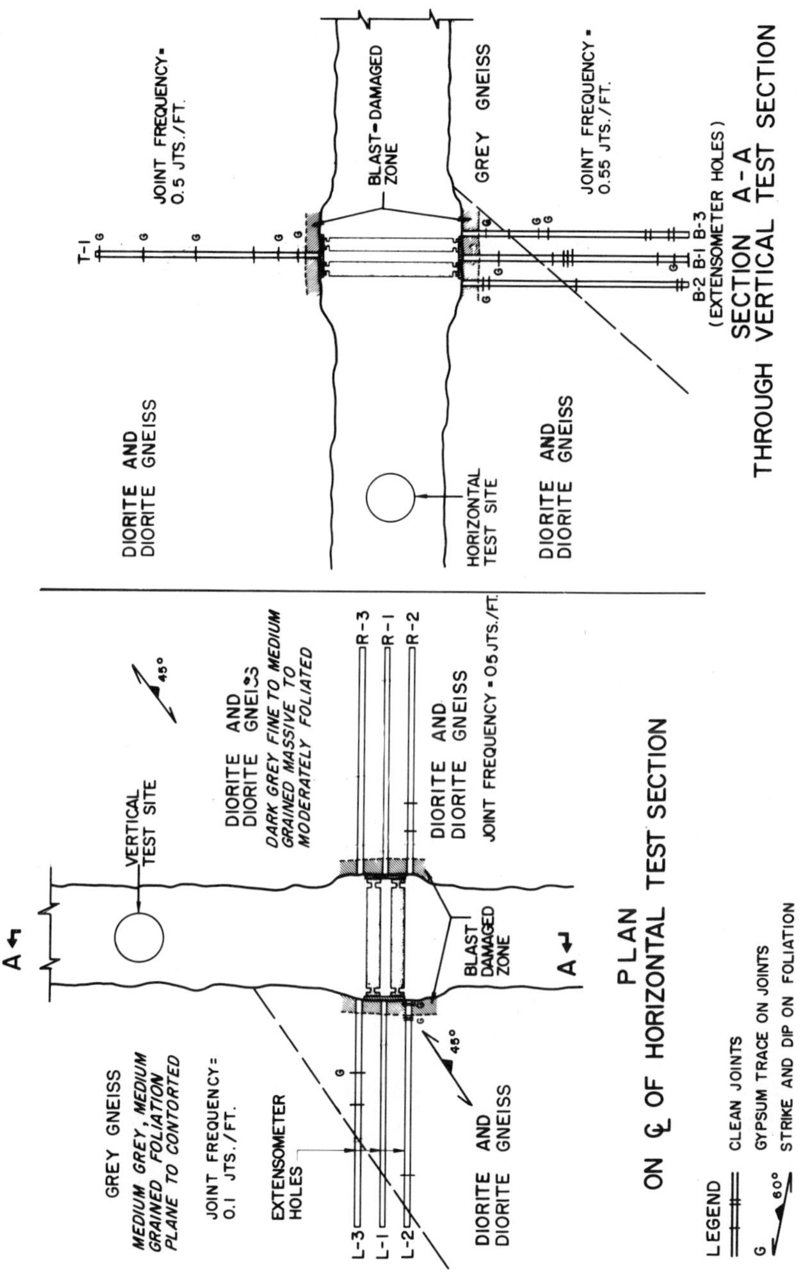

FIG. 6—*Test Drift 3 jacking sites—rock quality.*

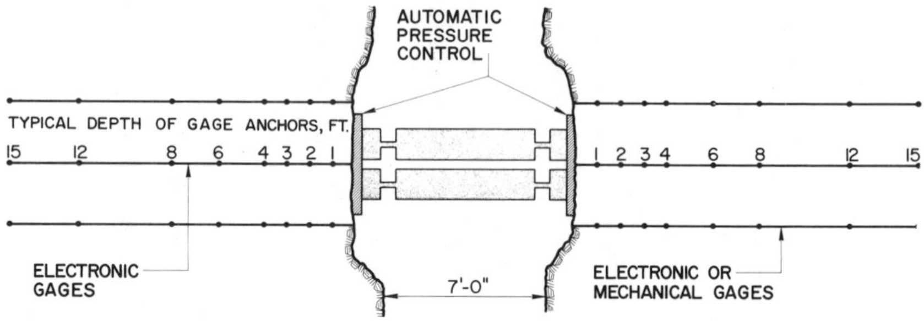

FIG. 7—*Typical test setup.*

end of each column to mechanically adjust the equipment, and wooden chipboards to provide a seat between the flatjack and the aluminum head.

Thrust against the rock was provided by the expansion of the flatjacks with oil under pressure from an automatic, air-driven stabilizing pump. A pressure transducer was provided in the lines to measure oil pressure, and an automatic pressure recorder was used to provide pressure surveillance.

Contact between the rock and the flatjack was made with hand-placed mortar. The mortar was mixed to a thick, plastic consistency and impressed into the space. The extensometer head was isolated from the mortar by oakum, which was later removed. The flatjacks were smeared with oil to prevent bonding with the mortar, and the test was left for two days to allow the mortar to harden.

Electrical borehole extensometers were then installed with a wire tension of approximately 30 lb, and mechanical extensometers were also placed at the edge of the plate jack with a wire tension of 15 lb. Central mechanical gages were installed to measure the diametrical deformation of the drift.

Test Procedure

The pressure was increased to the maximum in increments of 250 psi, and all gages were read at each increment. The maximum pressure varied for the tests from 1000 to 1350 psi. At 500 psi, 1000 psi, and maximum, the pressure was maintained until the deformations had stabilized, and at each of these pressures the loading was recycled through zero with deformation readings taken at each 250-psi increment. Each test took about three weeks to complete.

After this general procedure was completed, the load was cycled several times from zero to maximum pressure, deformation readings being taken at each 250-psi increment. Deformation readings on the electrical extensometers were taken with a null-balancing bridge. Mechanical gages sensitive to 0.001 and 0.0001 in. were used on the remaining gages. Air temperature at the jack was also recorded, although it was observed to be constant within the drill holes and was not considered to be a significant variable.

Results and Interpretation of Tests

Mathematical Models

Two mathematical techniques were used to model the elastic stress-strain behavior of the rock beneath the flatjack. These models were then used for correlation with the measured deformations so that the moduli could be evaluated.

The first method utilized the standard Bouissinesq equation for uniform loading of an elastic half space; but as the area loaded was actually "doughnut" shaped, the equation had been modified accordingly. The following equation was used to calculate the modulus at any depth beneath the center line of the jack:

$$E = \frac{q\,(1+u)}{\delta}\left[\frac{z^2}{\sqrt{z^2+a_1^{\,2}}} - \frac{z^2}{\sqrt{z^2+a_2^{\,2}}} + 2\,(1-u)\sqrt{z^2+a_2^{\,2}} - 2(1-u)\sqrt{z^2+a_1^{\,2}}\right] \quad \ldots\ldots(1)$$

where:

δ = deflection of point, in., where modulus E is required, psi;
q = unit load on plate jack, psi;
u = Poisson's ratio = 0.25;
z = depth below plate jack, in.;
a_2 = external radius of plate jack, in.; and
a_1 = internal radius of plate jack, in.

The second method utilized three-dimensional finite elements to model the strain response at any point beneath the plate jack. The solution is constrained to symmetrical loading, which is coincident with the boundary conditions of the plate jack test. This axisymmetric solution is simple, flexible, and can be easily altered to simulate the observed deformation response of the rock, thereby modelling the variations in elastic moduli with depth. This technique also simplifies the interpretation of measurements taken at any point behind the plate jack. Figure 8 shows a typical deformation plot for a section taken through the center line of the plate jack.

Analysis of Data

Both the total and the recoverable deformations measured for each

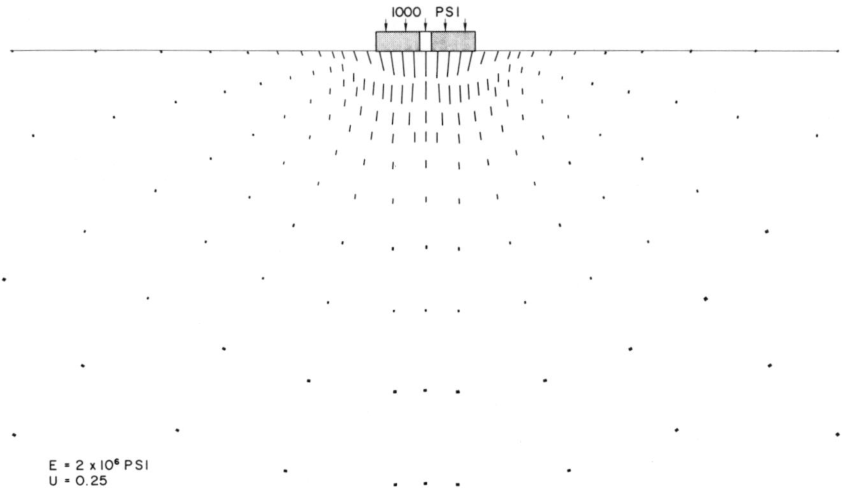

FIG. 8—*Axisymmetric displacements.*

increment of load have been used to calculate the moduli for the rock mass. Separate calculations were performed for the mechanical gages and for the electrical gages to ascertain the compatibility of the results. Three mathematical techniques were used:

1. The Boussinesq solution assuming no change in elastic constants with depth.

2. The axisymmetric solution assuming no change in the elastic constants from the rock surface to the anchor point used in the calculation.

3. The axisymmetric solution assuming increasing moduli with depth. (The function was fitted to the information calculated from 1 and 2.)

Figures 9 to 13 show all of the significant data obtained from the mechanical and electrical gages. For a homogeneous, isotropic material which responds elastically, the data should yield smooth functions. The actual deformations indicated a deviance from this ideal response; however, as the trend of data was in compliance with the theoretical response, it was judged that realistic moduli would be derived from an averaging technique. Accordingly, smooth curves were fitted to the data.

Figure 14 shows a comparison between a theoretical curve, which would apply to a material having a constant modulus, and the observed curve for that test. In every case there was a marked dissimilarity in shape near the surface, which indicated that rock near the surface of the drift was of markedly lower modulus than rock at depth. As the axisymmetric finite element technique is very flexible, it was adjusted to simulate material with a modulus increasing with depth. The deformation curve result-

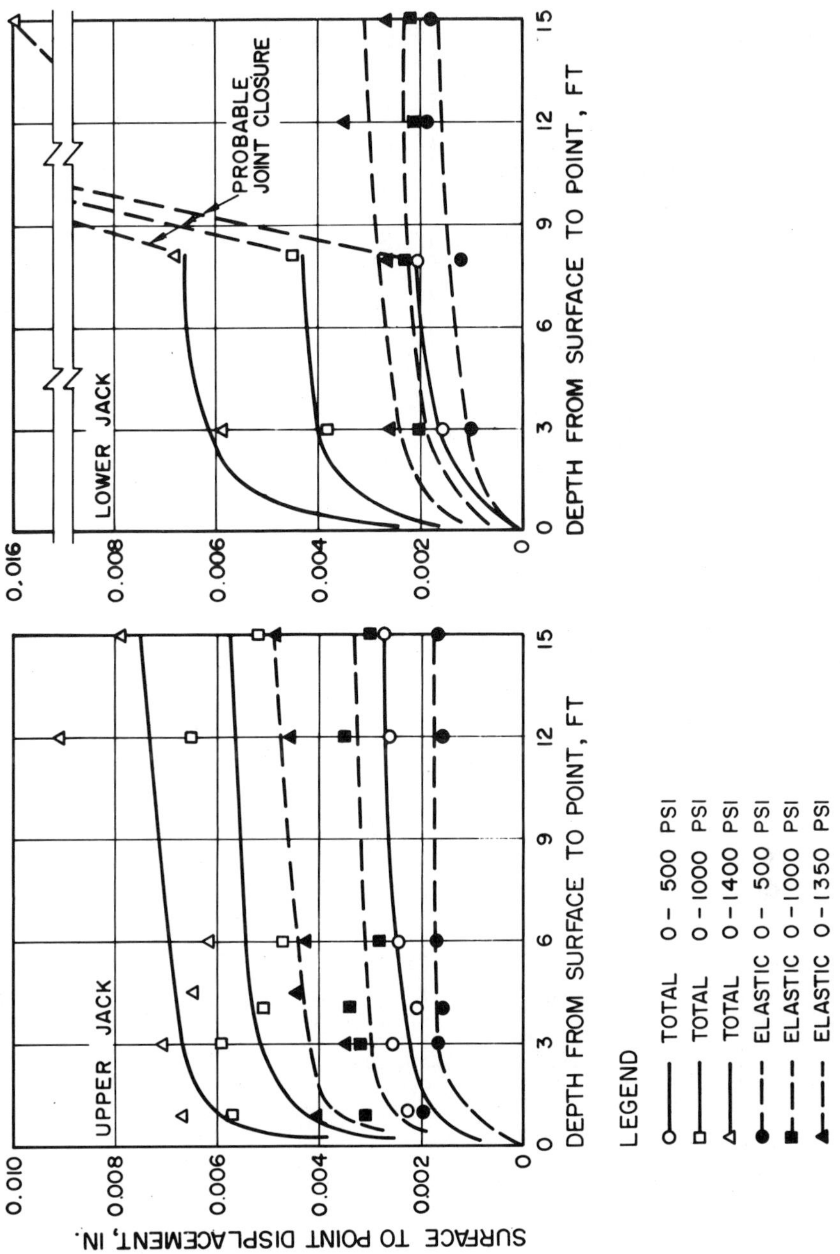

FIG. 9—*Test 3V results.*

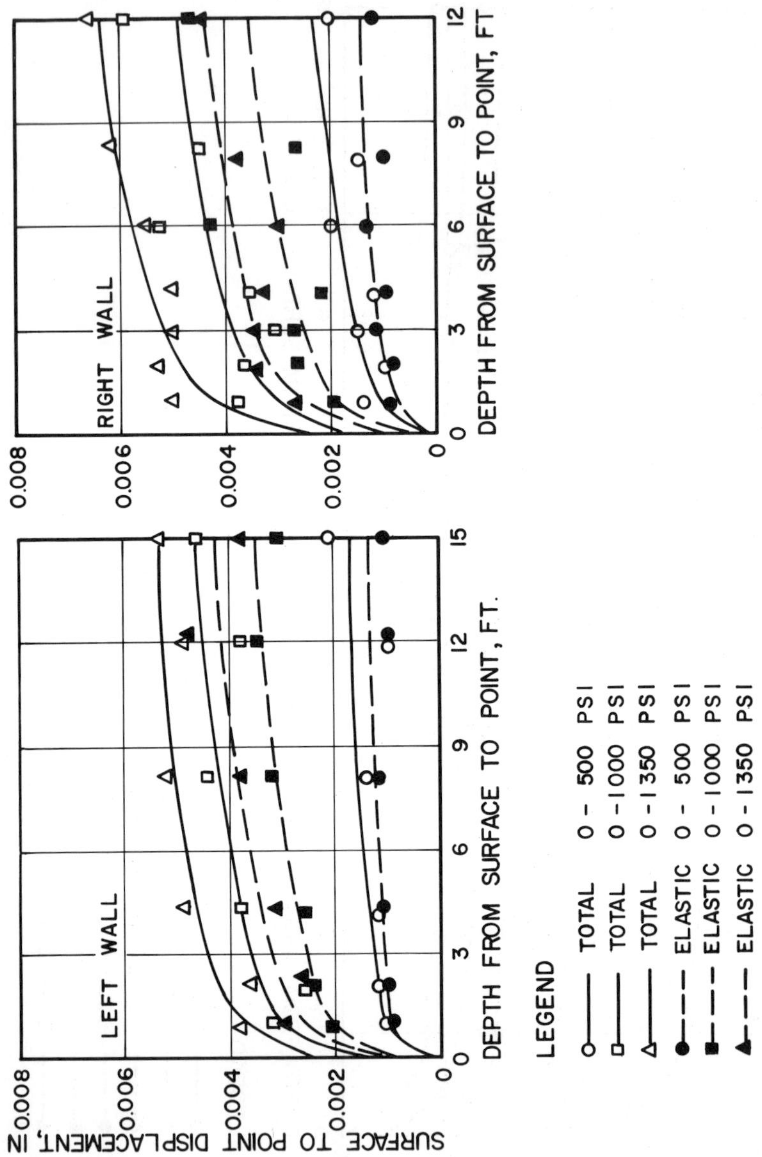

FIG. 10—*Test 2H results.*

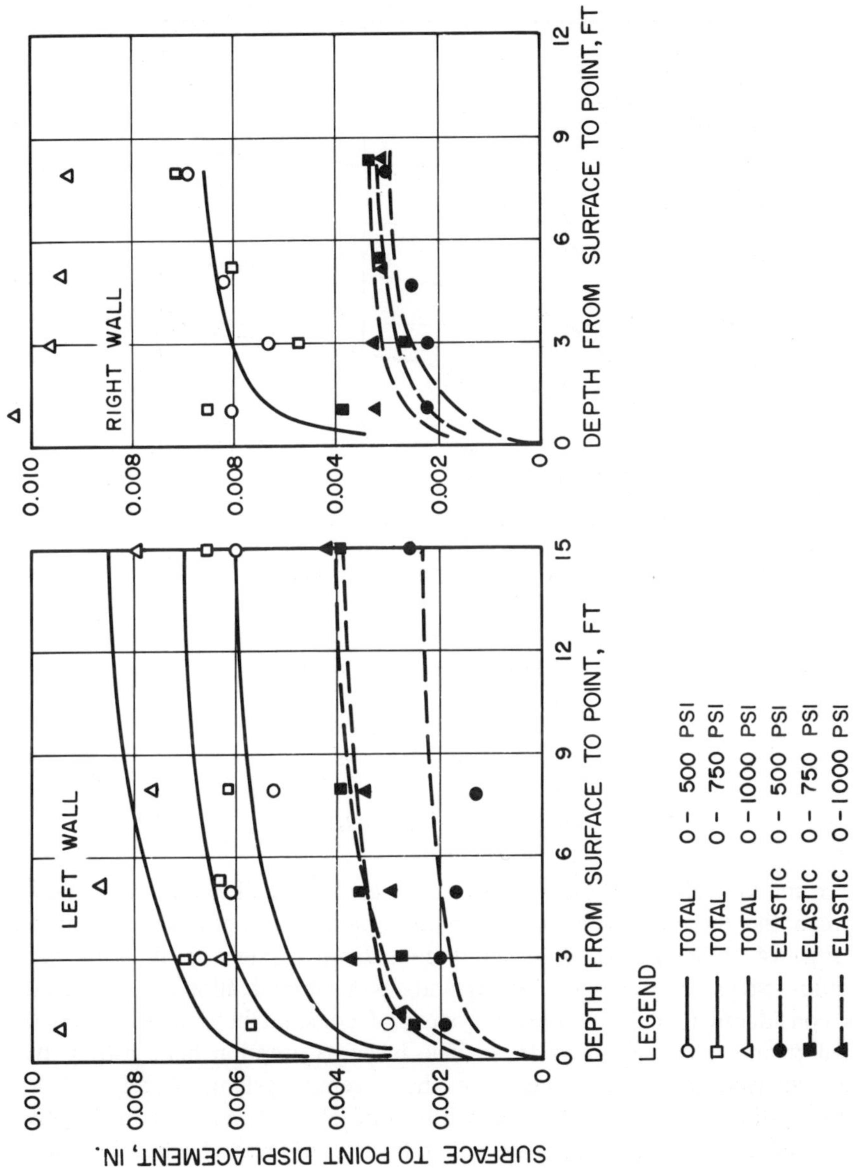

FIG. 11—*Test 3H results.*

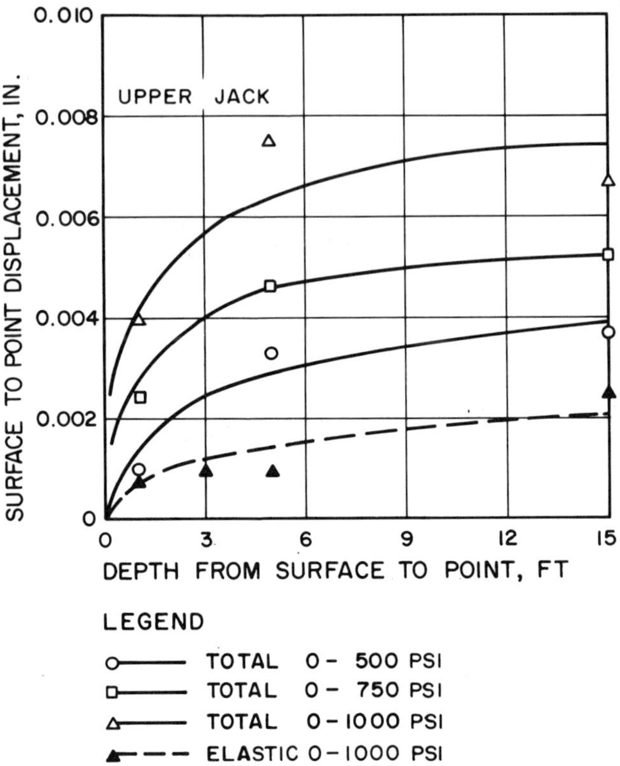

FIG. 12—*Test 1V results.*

ing from equivalent stresses was then calculated, and better agreement between the theoretical and observed deflections was obtained, as shown on Fig. 14. Nonetheless, it was not considered feasible to fit a precise modulus curve to each set of data; and, as a simplification, the material was considered to be a two-layer system of constant modulus between 0 and 3 ft and between 3 and 15 ft. The basic assumption here is that the stress distribution is unaffected by the layered system; therefore, the resulting differential displacements within one of the layers remain largely unaffected by the presence of the other layer. This was checked by comparing a two-layer system with a homogeneous material. The comparison showed that the differential deformations between any two points within a layer of constant modulus were largely independent of the existence of the other layer. Table 1 was then prepared by fitting the displacements within the axisymmetric model shown in Fig. 8.

Modulus values have also been calculated from the displacements measured by the internal dial gages—which measure only surface deformation

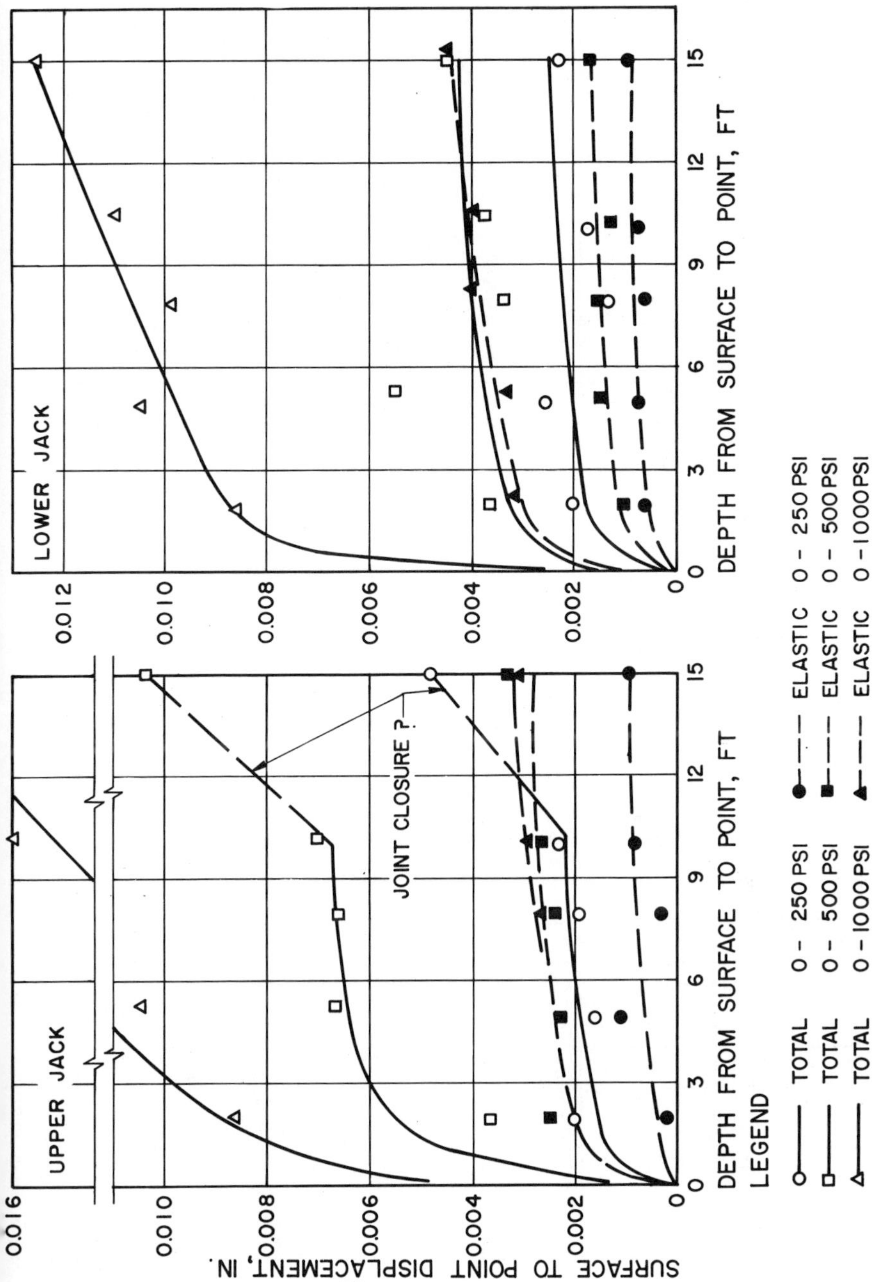

FIG. 13—*Test 2V results.*

TABLE 1—*Modulus values determined by the axisymmetric model.*

0 – 250 P.S.I. TEST No.

Depth	Quantity	1-V TOP	1-V BTM	2-V TOP	2-V BTM	2-H LEFT	2-H RT.	3-V TOP	3-V BTM	3-H LEFT	3-H RT.
0' TO 3'-0"	MODULUS FOR TOTAL (ELASTIC & INELASTIC) DEFORMATION (P.S.I. X 10^6)			2·1	2·0						
	ELASTIC MODULUS (P.S.I. X 10^6)			9·0	6·6						
	RATIO: ELASTIC DEFORMATION / TOTAL DEFORMATION			·23	·30						
3'-0" TO 15'-0"	MODULUS FOR TOTAL (ELASTIC & INELASTIC) DEFORMATION (P.S.I. X 10^6)			0·81	3·8						
	ELASTIC MODULUS (P.S.I. X 10^6)			4·9	7·0						
	RATIO: ELASTIC DEFORMATION / TOTAL DEFORMATION			·16	·54						
0' TO 15'-0" (CENTRAL ELECTRONIC GAUGES)	MODULUS FOR TOTAL (ELASTIC & INELASTIC) DEFORMATION (P.S.I. X 10^6)			1·26	2·4						
	ELASTIC MODULUS (P.S.I. X 10^6)			6·7	6·7						
	RATIO: ELASTIC DEFORMATION / TOTAL DEFORMATION			·19	·36						
0' TO 15'-0" (CENTRAL DIAL GAUGE)	MODULUS FOR TOTAL (ELASTIC & INELASTIC) DEFORMATION (P.S.I. X 10^6)	3·0		1·5							
	ELASTIC MODULUS (P.S.I. X 10^6)										
	RATIO: ELASTIC DEFORMATION / TOTAL DEFORMATION										

0 – 500 P.S.I. TEST No.

Depth	Quantity	1-V TOP	1-V BTM	2-V TOP	2-V BTM	2-H LEFT	2-H RT.	3-V TOP	3-V BTM	3-H LEFT	3-H RT.
0' TO 3'-0"	MODULUS FOR TOTAL (ELASTIC & INELASTIC) DEFORMATION (P.S.I. X 10^6)	3·0		1·2	2·1	5·9	4·8	3·2	4·5	1·5	1·2
	ELASTIC MODULUS (P.S.I. X 10^6)			3·3	6·0	6·8	7·1	4·2	6·5	4·3	2·9
	RATIO: ELASTIC DEFORMATION / TOTAL DEFORMATION			·37	·33	·87	·67	·76	·69	·35	·42
3'-0" TO 15'-0"	MODULUS FOR TOTAL (ELASTIC & INELASTIC) DEFORMATION (P.S.I. X 10^6)	3·3		1·4	6·1	10·5	7·7	5·6	* 0·7	4·7	6·1
	ELASTIC MODULUS (P.S.I. X 10^6)			7·0	10·7	16·3	12·2	24·0	7·0	8·4	7·0
	RATIO: ELASTIC DEFORMATION / TOTAL DEFORMATION			·19	·56	·64	·63	·23	·58	·54	·87
0' TO 15'-0" (CENTRAL ELECTRONIC GAUGES)	MODULUS FOR TOTAL (ELASTIC & INELASTIC) DEFORMATION (P.S.I. X 10^6)	3·1		1·2	2·8	7·2	5·7	2·3	1·4	2·0	1·8
	ELASTIC MODULUS (P.S.I. X 10^6)			4·2	7·3	8·9	8·6	3·4	6·7	5·3	3·8
	RATIO: ELASTIC DEFORMATION / TOTAL DEFORMATION			·28	·39	·81	·66	·68	·20	·38	·47
0' TO 15'-0" (CENTRAL DIAL GAUGE)	MODULUS FOR TOTAL (ELASTIC & INELASTIC) DEFORMATION (P.S.I. X 10^6)	3·1		1·6		6·0		6·7		6·4	
	ELASTIC MODULUS (P.S.I. X 10^6)			9·0		9·7		17·3		13·5	
	RATIO: ELASTIC DEFORMATION / TOTAL DEFORMATION			·35		·63		·59		·53	

0 – 750 P.S.I. TEST No.

Depth	Quantity	1-V TOP	1-V BTM	2-V TOP	2-V BTM	2-H LEFT	2-H RT.	3-V TOP	3-V BTM	3-H LEFT	3-H RT.
0' TO 3'-0"	MODULUS FOR TOTAL (ELASTIC & INELASTIC) DEFORMATION (P.S.I. X 10^6)	2·7								1·8	1·8
	ELASTIC MODULUS (P.S.I. X 10^6)									3·5	3·4
	RATIO: ELASTIC DEFORMATION / TOTAL DEFORMATION									·52	·53
3'-0" TO 15'-0"	MODULUS FOR TOTAL (ELASTIC & INELASTIC) DEFORMATION (P.S.I. X 10^6)	5·6								8·2	5·2
	ELASTIC MODULUS (P.S.I. X 10^6)									8·9	14·5
	RATIO: ELASTIC DEFORMATION / TOTAL DEFORMATION									·93	·36
0' TO 15'-0" (CENTRAL ELECTRONIC GAUGES)	MODULUS FOR TOTAL (ELASTIC & INELASTIC) DEFORMATION (P.S.I. X 10^6)	3·4								2·7	2·5
	ELASTIC MODULUS (P.S.I. X 10^6)									4·6	4·9
	RATIO: ELASTIC DEFORMATION / TOTAL DEFORMATION									·57	·50
0' TO 15'-0" (CENTRAL DIAL GAUGE)	MODULUS FOR TOTAL (ELASTIC & INELASTIC) DEFORMATION (P.S.I. X 10^6)	3·8								5·6	
	ELASTIC MODULUS (P.S.I. X 10^6)									9·3	
	RATIO: ELASTIC DEFORMATION / TOTAL DEFORMATION									·60	

0 – 1000 P.S.I. TEST No.

Depth	Quantity	1-V TOP	1-V BTM	2-V TOP	2-V BTM	2-H LEFT	2-H RT.	3-V TOP	3-V BTM	3-H LEFT	3-H RT.
0' TO 3'-0"	MODULUS FOR TOTAL (ELASTIC & INELASTIC) DEFORMATION (P.S.I. X 10^6)	2·5		1·5	1·6	4·0	3·8	2·8	3·6	2·0	1·5
	ELASTIC MODULUS (P.S.I. X 10^6)	12·0			4·5	5·7	5·8	4·8	7·8	4·5	4·5
	RATIO: ELASTIC DEFORMATION / TOTAL DEFORMATION	·21			·35	·71	·66	·59	·46	·44	·33
3'-0" TO 15'-0"	MODULUS FOR TOTAL (ELASTIC & INELASTIC) DEFORMATION (P.S.I. X 10^6)	5·8			2·9	10·3	8·9	15·8	* 1·2	9·1	19·5
	ELASTIC MODULUS (P.S.I. X 10^6)	16·0			8·2	10·3	9·8	29·8	21·0	16·7	32·5
	RATIO: ELASTIC DEFORMATION / TOTAL DEFORMATION	·53			·35	1·0	·91	·53	·24	·55	·60
0' TO 15'-0" (CENTRAL ELECTRONIC GAUGES)	MODULUS FOR TOTAL (ELASTIC & INELASTIC) DEFORMATION (P.S.I. X 10^6)	3·3			1·9	5·3	4·9	4·2	2·0	2·9	2·4
	ELASTIC MODULUS (P.S.I. X 10^6)	11·5			5·5	6·9	6·9	7·3	10·5	6·4	7·9
	RATIO: ELASTIC DEFORMATION / TOTAL DEFORMATION	·40			·35	·77	·72	·58	·19	·34	·34
0' TO 15'-0" (CENTRAL DIAL GAUGE)	MODULUS FOR TOTAL (ELASTIC & INELASTIC) DEFORMATION (P.S.I. X 10^6)	4·0				4·6		4·5		4·1	
	ELASTIC MODULUS (P.S.I. X 10^6)	9·0				5·8		5·5		8·5	
	RATIO: ELASTIC DEFORMATION / TOTAL DEFORMATION	·54				·80		·82		·41	

0 – 1350 P.S.I. TEST No.

Depth	Quantity	1-V TOP	1-V BTM	2-V TOP	2-V BTM	2-H LEFT	2-H RT.	3-V TOP	3-V BTM	3-H LEFT	3-H RT.
0' TO 3'-0"	MODULUS FOR TOTAL (ELASTIC & INELASTIC) DEFORMATION (P.S.I. X 10^6)					4·4	3·7	2·9	3·2		
	ELASTIC MODULUS (P.S.I. X 10^6)					6·1	5·7	4·7	8·2		
	RATIO: ELASTIC DEFORMATION / TOTAL DEFORMATION					·73	·66	·61	·40		
3'-0" TO 15'-0"	MODULUS FOR TOTAL (ELASTIC & INELASTIC) DEFORMATION (P.S.I. X 10^6)					14·0	10·7	16·0	* 1·4		
	ELASTIC MODULUS (P.S.I. X 10^6)					14·0	13·2	17·4	19·2		
	RATIO: ELASTIC DEFORMATION / TOTAL DEFORMATION					1·0	·82	·91	·25		
0' TO 15'-0" (CENTRAL ELECTRONIC GAUGES)	MODULUS FOR TOTAL (ELASTIC & INELASTIC) DEFORMATION (P.S.I. X 10^6)					6·0	5·1	4·4	2·1		
	ELASTIC MODULUS (P.S.I. X 10^6)					7·9	7·4	6·7	10·5		
	RATIO: ELASTIC DEFORMATION / TOTAL DEFORMATION					·78	·69	·65	·20		
0' TO 15'-0" (CENTRAL DIAL GAUGE)	MODULUS FOR TOTAL (ELASTIC & INELASTIC) DEFORMATION (P.S.I. X 10^6)					4·8		3·7			
	ELASTIC MODULUS (P.S.I. X 10^6)					6·6		6·1			
	RATIO: ELASTIC DEFORMATION / TOTAL DEFORMATION					·74		·62			

NOTE:
* – INDICATES JOINT

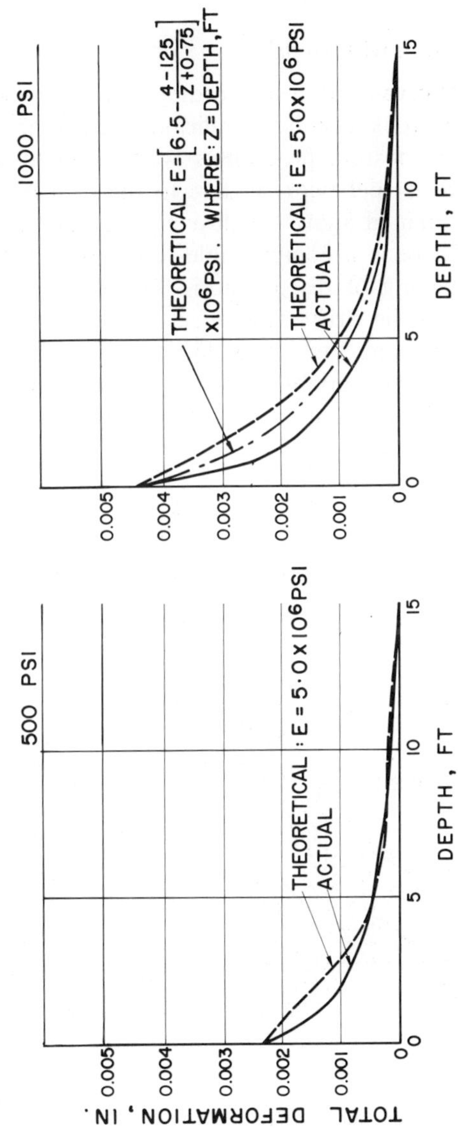

FIG. 14—*Theoretical versus actual deformation.*

of the rock—and from the total displacement measured by the electronical gages between the rock surface and the deepest anchor. The calculations assumed a constant modulus with depth, and the results are shown in Table 1.

Modulus Variation with Depth

Other investigators, such as Cording [*3*] have shown that the modulus of near-surface rock can be significantly less than the modulus of mass rock. Therefore, testing procedures were adopted to delineate this zone and to observe its behavior relative to the deeper rock.

The data obtained show conclusively that the modulus is not constant with depth. There is a compressible layer around the test drift, probably due to the opening of discontinuities by blasting, which allows inelastic strain relief into the drift. This effect is clearly shown in Fig. 15, where the deformation from 0 to 1 ft is 33 percent of the total deformation. The compressible layer is variable in depth and depends on the amount of blast damage and rock preparation prior to testing and on the magnitude of the

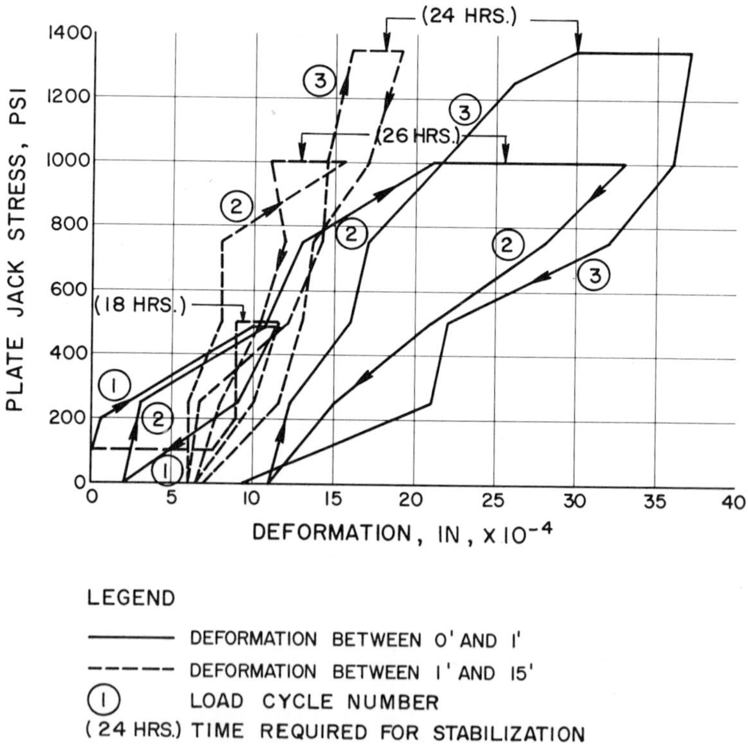

FIG. 15—*Load-deformation curves.*

stresses relative to the strength of the mass material. For the tests at Churchill Falls, it is believed that the layer was less than 3 ft thick and that the major deformation was contained within 1 ft. The moduli of mass rock calculated from the total observed deformations can be thus completely overshadowed by variations within this layer.

Within the compressible layer deformation moduli as low as 1.0×10^6 psi were calculated assuming a constant modulus from 0 to 3 ft. If the depth of the layer is varied, the modulus will also vary and can be as low as 0.5×10^6 psi for the range from 0 to one ft. For depths beyond the compressible layer, the deformation modulus does reflect minor inelastic movement and is correspondingly less than the elastic modulus. It is difficult, however, to determine a single value, and, for purposes of predicting mass rock behavior, the deformation modulus has been established as 6×10^6 psi.

Inelastic Behavior

A detailed study of load-deformation curves indicates that no appreciable inelastic set is observed beyond 2 to 3 ft, which is compatible with the observed rock quality (Fig. 16).

Irreversible deformations equal to 80 percent of the total deformation

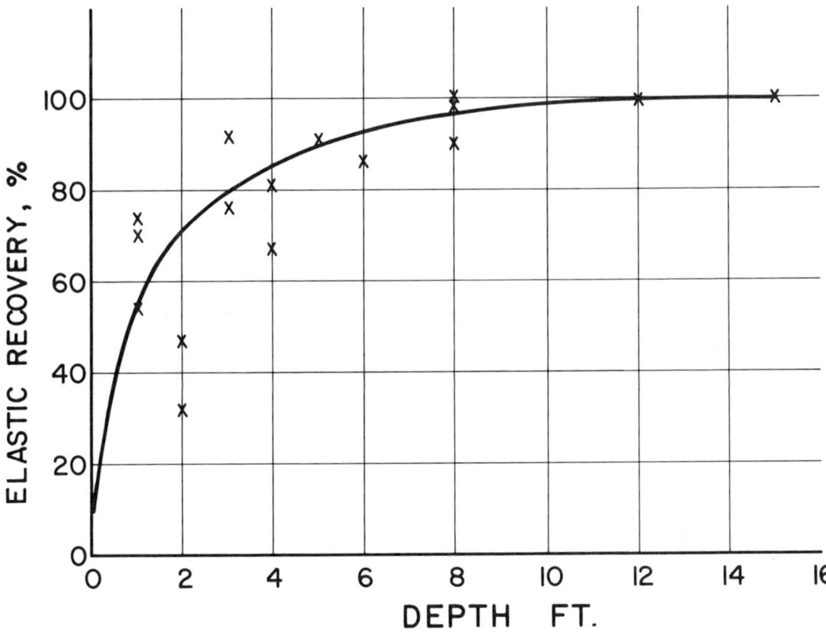

FIG. 16—*Inelastic deformation.*

were observed in the compressible layer, although they were generally less than 60 percent. This "inelastic set" is primarily attributed to the first cycle of loading. Subsequent cycles show almost fully recoverable deformations, as is shown in Fig. 15. Although the ratio of elastic to total deformation for depths beyond 3 ft (Table 1) indicates inelastic set in the rock, this is greatly influenced at greater depths by data interpretation and testing errors. The mass rock beyond the compressible layer has a high modulus; hence, the observed deformations are very small, and minor variations in the deformations greatly affect this ratio. In the compressible layer the deformations are much greater and can absorb these variations without having a significant effect on the ratio.

Closure of joints or cracks has been observed and can be reliably fixed between anchor points, for example, Fig. 9 indicates a discontinuous movement between 8 and 12 ft. As a result, the deformation modulus is low, whereas the elastic modulus is unaffected because the crack is closed under the initial loading. Commonly, this effect is limited to the compressible layer and is not considered to be a factor for the rock at depth. It is important to have a number of anchor points to delineate the position of any open discontinuity, since the presence of such a feature markedly affects the calculated modulus.

Inelastic set in the rock mass does not necessarily mean that fractures have been permanently closed. This is apparent from a study of the elastic moduli of near-surface rock, where the elastic modulus obtained by cycling is much lower than the modulus of deeper, intact rock. Although loading closes the cracks, they also partially reopen when the stress is removed. Continued cyclic loading results in a consistent and repeatable hysteresis loop. This may be due to the tangential stress around the periphery of the opening, which compresses layers of rock formed by stress relief or blast damage so that they tend to act somewhat like leaf springs. The effect also includes a time delay, and a significant stress may be required to reclose the cracks. Reloading curves, therefore, do not necessarily pass through the same point of deformation that was reached on the initial load cycle, as shown in Fig. 15. However, if the load is maintained at a given stress level, the previous deformation will be obtained. This viscous effect is considered to be almost wholly the result of crack closure and movement along discontinuities.

These inelastic effects can have a marked influence on the rock behavior at any stress level. They are particularly important when dealing with the design of lined pressure conduits. Carefully controlled grouting can largely remove the inelastic behavior of the near-surface rock by filling and prestressing the open cracks with grout. Some investigators [5] have shown improvements in the deformation modulus by a factor of 3.75, though it is commonly less than 2. It is expected that grouting of the rock surrounding the penstocks at Churchill Falls will produce similar beneficial results.

Vertical Tests versus Horizontal Tests

Previous investigators have shown that the deformation modulus can be influenced by the direction of testing. For the Churchill Falls rock, there are persistent, near-horizontal joints which are locally filled with soft gypsum up to ⅛ in. thick, although it most commonly appears as a thin film. These joints are much more numerous than the inclined or vertical joints encountered in the horizontal drift holes, as is shown on Figs. 5 and 6. It was therefore expected that the vertical tests would yield a lower modulus due to the more pronounced relaxation of the rock in the vertical direction and that perhaps they would also reflect the compression of the horizontal, gypsum-coated joints. There is evidence of this in Drift 2 where the vertical deformation moduli are approximately one half of the horizontal values. The effect was masked in the near-surface layer in Drift 3 due to the large movements although it was suggested at depths beyond 3 ft.

In the areas tested the foliation was moderate to indistinct. As all the vertical and horizontal tests were approximately parallel to the foliation, its overall effect on the testing was not established. It is judged, however, that the data would not reflect with any certainty any effect of the foliation.

Modulus versus Stress Level

Other investigators [*6*] have commonly shown that the modulus increases as the applied stress is increased. This is assumed to result from the closure of cracks or joints in the mass rock under low stresses, making the material stiffer at higher stresses. The same effect was observed with these tests although some of the data did not follow this pattern. The observed rock movement was very complex and variable, with discontinuous and viscous-like movement occurring under different stress levels. Consequently, the deformation modulus can be sharply reduced when the applied stress closes an open discontinuity or causes inelastic slip along a joint.

It is difficult to relate the modulus of the mass rock to the applied stress because the rapid stress decay beneath the jack results in low stresses being generated at depth. Hence, where stress changes for a prototype structure are large, extrapolation of modulus values for use in predicting mass rock behavior becomes very difficult. A promising method has been outlined by Cording in which rock quality at a given point is weighted relative to the applied stress. In general, with the homogeneous, high-quality rock at Churchill Falls, the method does not alter the modulus values. It is considered that the modulus of deformation for this quality of rock is relatively independent of the applied stress level.

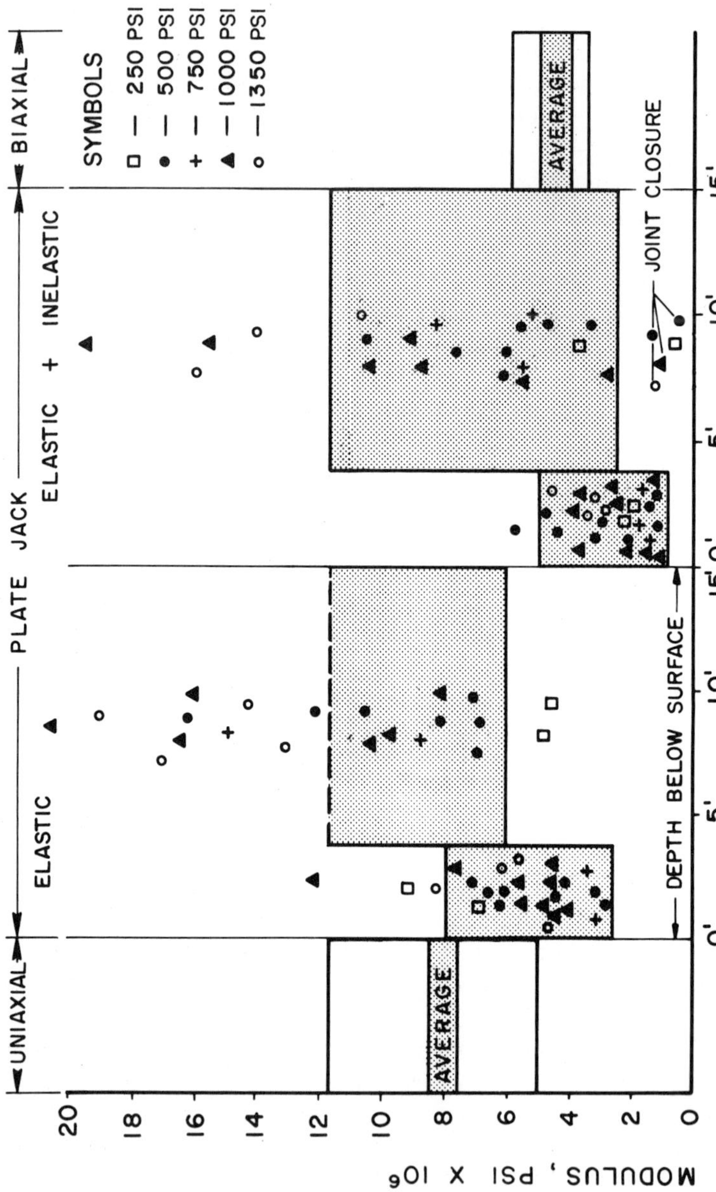

FIG. 17—*Comparison of moduli.*

Figure 17 shows the various modulus values that have been obtained by the different test methods. The uniaxial tests were performed on NX specimens in the laboratory. The biaxial tests were performed with a hydrostatic compressive load around intact 6-in.-diameter specimens containing a central EX hole, within which the radial deformations were determined. These specimens were obtained from overcoring tests.

There is a wide variation in the computed moduli, depending upon the type of test performed. Although the biaxial tests were conducted on intact specimens, the modulus values are approximately 50 percent of those calculated from the NX specimens. Triaxial confinement of rock is obtained beneath the plate jacks, and the elastic moduli are correspondingly high; unfortunately, data scatter militates against comparative conclusions. It is believed that microfissures may be formed around the EX hole in each overcored specimen. As the behavior of rock is largely controlled by inherent discontinuities and the applied stress field, the biaxial test leads to larger radial deformations that result in a lower modulus than would be obtained from an undamaged portion of the same specimen. This is a very significant factor in calculating *in situ* stresses from overcoring tests. It is important, therefore, to relate the material and its expected mode of behavior to the most significant type of test.

Limitations of Tests

Mathematical Models

To calculate rock modulus values from plate jack test results, it is necessary to choose a set of stress distribution equations or to provide some other mathematical model as a basis of comparison. Classically, the approach has entailed the use of Boussinesq equations for the solution of the stress values within an elastic half-space under the application of a point load. Simple integration extends the equations to include the case of an applied "doughnut-shaped" load. From the stress field solution the field modulus necessary to provide any given displacement can be calculated. With this model the inherent assumptions of reversible elasticity and isotropy incur errors when compared with reality. The greatest sources of error with the assumption of isotropy are that elasticity prevails and that the modulus remains constant with depth.

A more flexible basis for calculation of rock moduli lies in the use of an axisymmetric solution utilizing finite element analysis. This method allows the rock system to be layered or to exhibit a varying modulus with depth, but there is a limitation in the necessity of dealing with a finite field having certain boundary conditions. For this particular case, a boundary 15 ft from the surface was fixed on rollers, generating an error in the total deflection of the surface. As this error was a cumulative result, the differential deflection between any two points within the field was negligible. This method can be expected to be a more suitable mathematical technique for modelling rock behavior since inelasticity is more easily handled. However, the complex response of mass rock under changing stress environments requires further study before truly representative models will be found.

Instrumentation

The calculated modulus for rock at depth was highly sensitive to experimental errors of measurement when the real modulus was greater than 5×10^6 psi. The relationship between the real and calculated modulus for a fixed experimental error of 0.0005 in. is outlined below. Note that the range diverges sharply as the real modulus increases.

Modulus for Interval of 3 to 15 Ft (Load, 500 psi)

Real	*Calculated*
5×10^6 psi	$3.3–10 \times 10^6$ psi
6	3.7–15
7	4.1–23
9	4.7–83
10	5.0–∞

It was evident that when dealing with the deeper-lying, high modulus rock, the precision and accuracy of measurements becomes of paramount importance. The low stresses generated in this region, coupled with the high modulus, result in very small deflections, thus placing increasing emphasis on experimental errors. The situation is aggravated by the increasing possibility of error for deeper-lying anchor points when all measurements are taken relative to the surface. Thus, it is desirable to have a system which measures directly between any two points at any depth. When it is necessary to analyze a deeper-lying layer in high-quality rock, it is essential that measuring techniques be correspondingly accurate or that the applied stress level be increased sufficiently such that experimental errors remain insignificant with respect to the generated displacements.

Conclusions

1. The near-surface compressible layer of blast-damaged rock can dominate the overall deformation response in plate jack testing, especially when the undamaged rock has a high modulus. Precise and accurate measurements are required to define rock behavior, especially at depths where low stress changes are generated. Instrumentation must, therefore, be specifically selected to accurately measure the response of the rock which is to be tested. It is equally important that sufficient data be obtained to

define the trends of behavior as data scatter can be diverse and difficult to interpret.

2. Results of the plate jack testing were compatible with the excellent rock quality observed at Churchill Falls. Although deformation moduli as low as 500,000 psi were measured within a depth of 1 ft, beyond 1 ft the modulus increased sharply and indicated a value of 5 to 7 $\times$ 10^6 psi. Moduli obtained from cyclic loading varied from 2 $\times$ 10^6 psi within 1 ft, to a value at depth which was compatible with intact laboratory specimens.

3. The mass rock at Churchill Falls is hard and durable, and, in general, no appreciable irreversible movement was indicated for depths beyond 2 to 3 ft. Inelastic set generally less than 60 percent was observed in the compressible near-surface layer. This occurred under the first cycle of loading, with subsequent cycles showing almost fully recoverable deformations.

4. Vertical tests indicated a lower deformation modulus than did horizontal tests. This is attributed to the more numerous horizontal joints and may reflect some effect of the gypsum coating commonly associated with this joint set.

5. It is important that a flexible and adequate mathematical model be used to provide correlation with any observed behavior. The elastic models used here provided a fairly crude tool for correlation, but the more adaptable computer techniques provide promise for future work.

6. Results of modulus testing of rock reflect the physical makeup and size of the specimen, as well as the *in situ* and applied stress environment. It is important to relate the material and its expected mode of behavior to the most significant type of test so that the calculated modulus is meaningful.

Acknowledgments

The development is owned by Churchill Falls (Labrador) Corp., Ltd., (CFLCo). Design and construction management are the responsibility of the joint venture, Acres Canadian Bechtel (ACB). The writers wish to express their thanks to D. Wermenlinger, manager of engineering (CFLCo), and J. G. S. Thomson, manager of engineering (ACB), for their interest in the work and approval of this paper. R. J. Conlon, H. G. Acres Co., Ltd., and D. U. Deere, special consultant, provided constructive comments throughout the study and during the preparation of the paper. A. H. Merritt, J. J. Chen, and T. Neff were largely responsible for the field testing. Special thanks are due B. Sellers, Terrametrics Inc., for initial direction of the field work.

References

[1] Dodd, S. D., "Morrow Point Underground Powerplant Rock Mechanics In-

vestigations," Water Resources Technical Publication, United States Bureau of Reclamation, March 1967.

[2] Alexander, L. G., Worotnicki, G. and Aubrey, K., "Stress and Deformation in Rock Support, Tumut 1 and 2 Underground Power Stations," 4th Australia-New Zealand Conference on Soil Mechanics and Foundation Engineering, 1963.

[3] Cording, E. J., "The Stability during Construction of Three Large Underground Openings in Rock," PhD thesis, University of Illinois, Urbana, Ill., 1966.

[4] Sigvaldason, O. T., Benson, R. P. and Thompson, M. V., "Stress Analysis of Underground Openings in Rock Using the Finite Element Method," presented to Engineering Institute of Canada, May 1968.

[5] Lancaster-Jones, P.F.F. in *Rock Mechanics in Engineering Practice,* John Wiley and Sons, New York, 1968.

[6] Monahan, C. J. and Sibley, E. A., "Rock Mechanics for Dworshak Dam," presented at the Annual Meeting, Geological Society of America, 1965.

[7] Deere, D. U. and Miller, R. P., "Engineering Classification and Index Properties for Intact Rock," Technical Report No. AFWL-TR-65-116, Air Force Weapons Laboratory, Kirtland Air Force Base, N. M., 1966.

E. J. Deklotz[1] *and B. P. Boisen*[1]

Development of Equipment for Determining Deformation Modulus and *In Situ* Stress By Means of Large Flatjacks

REFERENCE: Deklotz, E. J. and Boisen, B. P., **"Development of Equipment for Determining Deformation Modulus and *In Situ* Stress by Means of Large Flatjacks,"** *Determination of the* In Situ *Modulus of Deformation of Rock, ASTM STP 477,* American Society for Testing and Materials, 1970, pp. 117–125.

ABSTRACT: Development of equipment and proposed procedure for determining deformation modulus and *in situ* stress in rock masses is discussed. The procedure, which involves the use of large flatjacks, when perfected, will be considerably more economical than conventional, more elaborate methods such as the pressure chamber method, radial jacking, and the like. The flatjack procedure involves the cutting of a slot in the rock mass by means of a hydraulically powered 3-ft-diameter diamond saw, insertion of the large 3 by 6-ft flatjacks, and pressurizing of the jacks until cancellation is reached. The system has an advantage over conventional flatjack usage in that it permits pressurizing (testing) larger areas of rock, eliminates the need for grouting the flatjacks, and permits the determination of deformation modulus in addition to *in situ* stress.

KEY WORDS: stress analysis, modulus of elasticity, stresses, stress relieving, rock mechanics, rocks, construction equipment, tests, deformation, evaluation

In situ rock mass testing is laborious, expensive, and time consuming. This is particularly true for deformation modulus tests in which pressure chambers, radial jacks, or similar elaborate equipment must be assembled. A more economical procedure has recently been proposed.[2] This European

[1] Director and research engineer (1st Lt., U. S. Army CE), respectively, U. S. Army Corps of Engineers, Missouri River Division Laboratory, Omaha, Neb.

[2] Rocha, M., "Rock Mechanics in Portugal," *Proceedings of the First Congress of the International Society of Rock Mechanics,* 25 Sept.–1 Oct. 1966, Lisbon, Portugal, Vol. 3, 1967, p. 121.

method involves the use of large flatjacks, approximately 1 m wide and 2½ m long, inserted in slots previously cut in rock. Deformation of the rock mass is then measured at various pressure levels. In our work the method includes *in situ* stress determinations in addition to the deformation modulus. This paper, in the form of a progress report, discusses the equipment, initial field tests, and proposed test procedure.

The system described involves the use of a multiple-position caliper to monitor the four-step procedure of (1) the cutting a series of slots 3 ft long, ⅜ in. wide and 6 ft deep in the rock; (2) the insertion of flatjacks; (3) the measurement of deformation at various load levels, and (4) the determination of the cancellation pressure. The observations permit calculations of deformation modulus and *in situ* stress at that location and orientation. In addition to being considerably less expensive and time-consuming than conventional methods, the procedure presented here is particularly appealing because the usual flatjack grouting operation is elminated due to the resultant smooth, saw-cut rock surfaces.

The saw, which somewhat resembles a unicycle in principle, consists of a 3-ft-diameter diamond blade, a guide tube which houses sprockets and drive chains, and a 30-hp hydraulic motor at the top. It is supported in a truncated, pyramidal tubular steel tower which, in turn, rides on an adjustable positioning track. The track is securely anchored to the rock

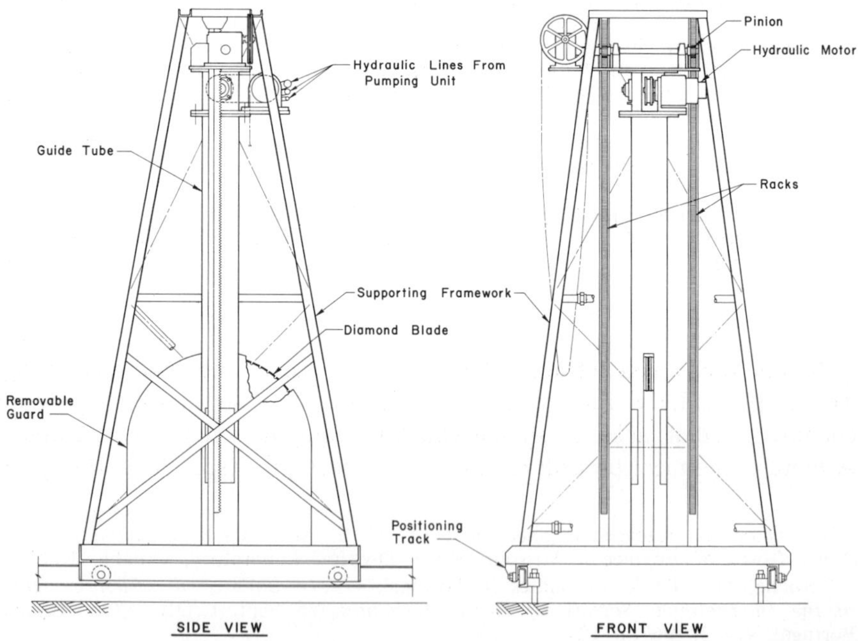

FIG. 1—*Rock saw.*

surface to provide the stability necessary during the sawing operation. Figure 1 is a schematic drawing of the saw, and Fig. 2 shows the layout of a typical test site.

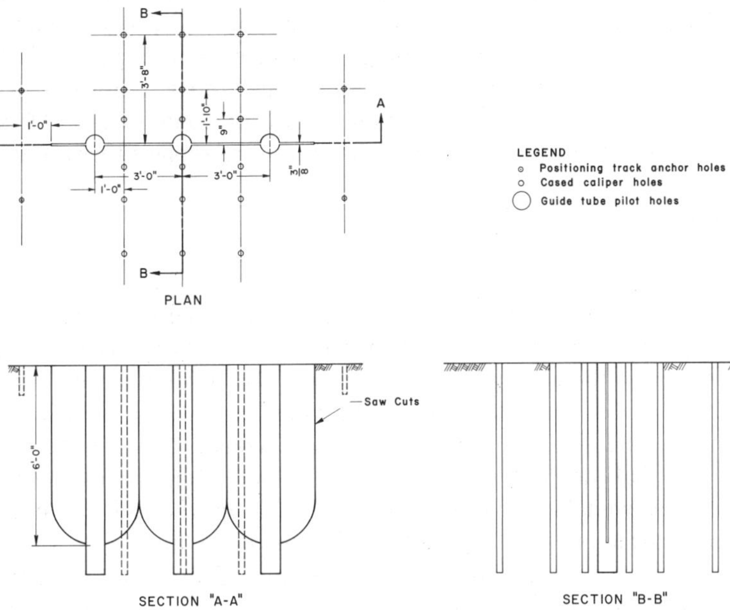

FIG. 2—*Test site hole layout.*

Test Procedure

Referring to Fig. 2, the proposed test procedure is as follows. A series of 2.36-in.-diameter (BX) holes is first drilled to a depth of 7 ft in a symmetrical pattern around the future saw cut. After the holes have been drilled, they are cased with plastic tubing (2-in. outside diameter by ⅛-in. wall), and the annular space around the tubing is packed with natural silica sand from Ottawa, Ill., graded to pass a No. 20 (841 μm) sieve and retain on a No. 30 (595 μm) sieve. Immediately after the casings are packed, readings are taken with a multiple-position caliper at 6-in. depth intervals between all pairs of cased holes. (The flatjack pressure required to reproduce these readings, after partial stress relief in the rock by the saw cut, is termed the cancellation pressure.) Three 7 ¾-in.-diameter holes are now drilled on 3-ft centers to act as pilot holes for the 7 ½-in.-diameter guide tube of the saw. Measurements with the calipers are taken again to determine if any relief has occurred as a result of the drilling of the pilot holes.

Following this, caliper readings are taken periodically until deformations due to stress relief have stabilized. When the area around the pilot holes has stabilized, the rock saw is assembled, and three adjoining saw cuts are made. Caliper readings are again taken until no further deformation can be detected. Semicylindrical grout plugs are now cast in the pilot holes on both sides of snug-fitting shims that have been previously placed in the saw cuts. The purpose of the plugs, cast in place in long plastic bags, is to fill the pilot holes so that the flatjacks can exert a uniform load over their full width without bursting from a lack of confinement. After the grout has reached adequate strength, the shims are replaced with flatjacks.

Equipment

Rock Saw

A hydraulic power system was selected to drive the sawblade, in preference to electric power, principally because of the excessive weight of an electric motor of the required 30-hp capacity. The hydraulic motor weighs only 60 lb and has about the same dimensions as a ⅓-hp electric motor. Furthermore, the hydraulic system offers a considerable economic advantage: in shakedown tests it was found that, although the hydraulic motor has an available maximum output of 30 hp, the saw can cut hard limestone satisfactorily at a preset maximum of 15 hp.

Roller chains transmit the output of the hydraulic motor to sprockets on both sides of the diamond blade. The sprockets are mounted on a two-part, 1 ½-in.-diameter shaft which, in turn, rotates two 7-in.-diameter flanges between which the diamond blade is mounted. Two shear pins pass through the flanges and the blade. Safety regulations established by industry specify a minimum flange diameter of 7 in. to provide blade stability during rotation. Thus the flange diameter dictates the minimum guide tube diameter.

The hydraulic motor, diamond blade, and guide tube are raised and lowered as a unit by means of a rack-and-pinion assembly. The pinions are mounted on a shaft coupled to the output side of a 20:1 worm gear reducer. The input shaft of the gear reducer is rotated by the operator by either a hand crank or a wheel and chain.

The racks are supported by 2-in.^{2} thin-walled steel tubes that are attached at both ends to the tubular steel tower. The tower base is such that it can be rolled on a positioning track in either direction parallel to the plane of the sawblade. The track, which is made of 4-in. steel channel, is anchored to the rock by grouted bolts.

The 36-in.-diameter diamond blades are of special design. The core is 0.250-in.-thick cold-rolled steel, and the cutting surface is 0.375 in. wide. The thicker-than-standard core is necessary because of the aluminum flange diameter; furthermore, it was considered desirable to increase the core thickness to guard against warpage, since the blade cuts with 43 percent of its circumference a majority of the time.

Water from a supply line within the guide tube is directed through twelve jets to each side of the blade to act as a coolant and to remove cuttings from the blade circumference. After the water has passed over the blade and drained into the pilot hole below the guide tube, it is removed by a suction line also mounted within the guide tube. This configuration reduces drag on the blade to a minimum, since it does not operate submerged in water and cuttings. Figure 3 is a photograph of the saw in operation.

FIG. 3—*Saw in operation.*

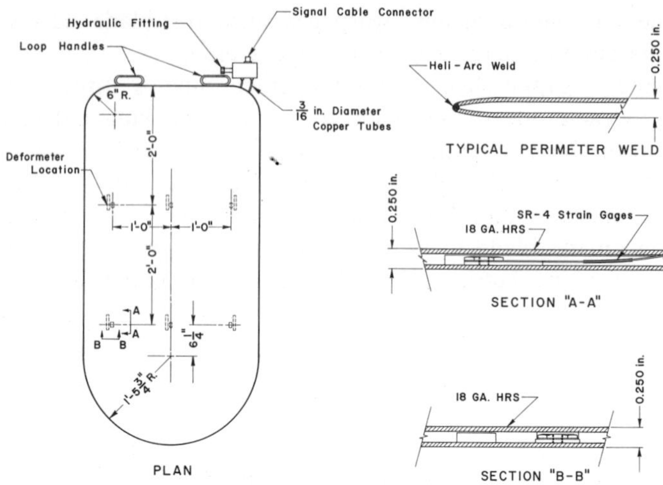

FIG. 4—*Details of flatjack construction.*

Flatjacks

Figure 4 shows the construction details of a flatjack. The jacks are made of 18-gauge hot-rolled mild steel. The perimeter is heli-arc welded with a stainless steel welding rod. Prior to completion of the perimeter weld, the copper tubes are silver-soldered in place, the deformeters mounted in the jacks, and the signal wires passed through the tubing and secured in the terminal block. An array of deformeters, or rock deformation sensing devices, is mounted in each flatjack. Each deformeter consists of two SR-4 stain gages cemented with epoxy resin to opposite sides of a 2 ½-in.-long, curved, spring-like strip of hardened beryllium-copper alloy. The strain gages are wired in a half-bridge circuit, and each deformeter is monitored independently. Prior to mounting in the flatjacks, the deformeters are calibrated for deformation versus indicator output and for hydraulic pressure versus indicator output. Hydraulic pressure of the magnitude involved has an influence on the strain gages and must be accounted for in the interpretation of deformeter readings. The deformeters are calibrated for the effect of these pressures in both the flat and curved positions, should there be any difference. Figure 5 shows a typical deformeter calibration curve.

The flatjack design evolved from a series of prototypes which were evaluated for their ability to maintain and respond to pressure. Response was determined by loading in a compression testing machine, the results of which are shown in Fig. 6. It should be observed that at pressures greater than 1500 psi, the area of the jack affected by the perimeter weld is of little significance, that is, almost total response is attained above this pressure level.

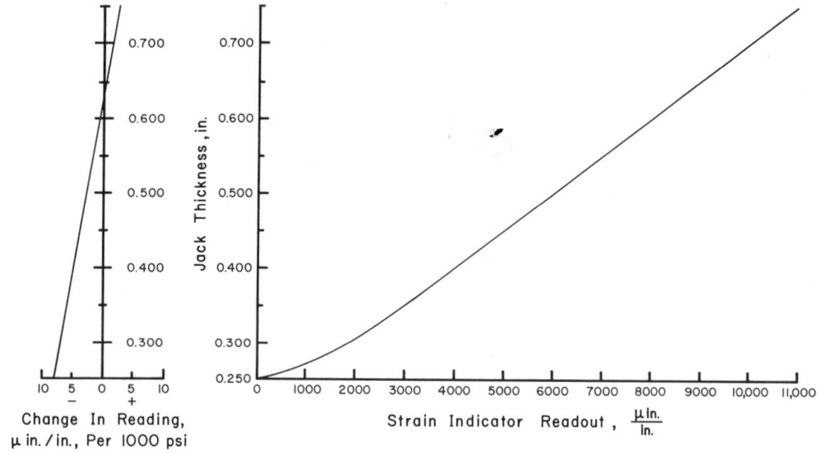

FIG. 5—*Typical deformeter calibration curve.*

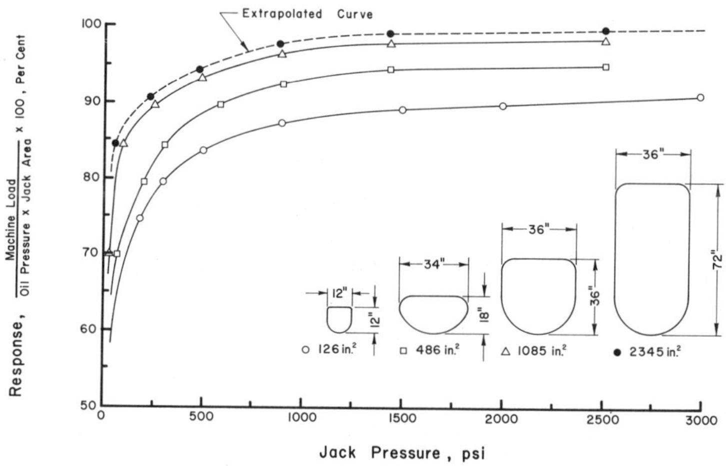

FIG. 6—*Flatjack response.*

Multiple-Position Caliper

Rock deformation is monitored by means of a multiple-position caliper at various depths and distances from the saw cut. The caliper measures distances between the cased drill holes symmetrically spaced on either side of the saw cuts. The caliper, Fig. 7, consists of two rods that can be positioned for various span lengths on the indexing bar. A self-tightening anchor is at the end of one rod and a calibrated sensor is at the end of the other. The caliper rods are supported in the hole at the casing collar. Laboratory tests show that the caliper has an accuracy of 0.001 in. and

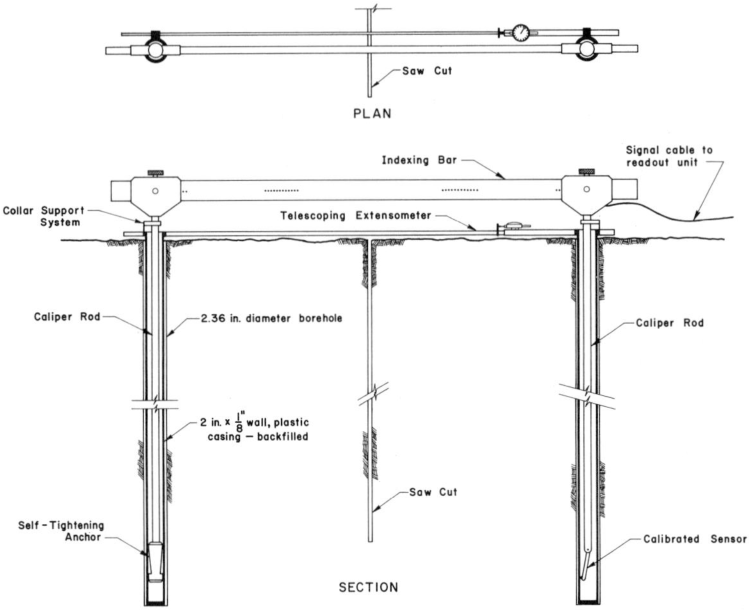

FIG. 7—*Multiple position caliper.*

that readings are reproducible to within 0.0005 in. over a range of 14 to 48 in. The caliper is in the prototype stage; future modifications will include the use of invar steel and an increase in the operational depth to over 6 ft. Due to the top-heavy nature of the caliper, readings taken within 6 in. of the casing collars are not reliable. Therefore, a telescoping extensometer is used to measure collar-to-collar displacements.

Field Tests

Field trials to date have been limited to procedural development and performance testing of the saw and appurtenant equipment. The tests were made in a hard limestone in which an average cutting rate of 0.5 in./min was maintained. This provided a full 6-ft-deep cut in approximately 2½ h. The importance of good alignment of all the holes, the pilot, and the caliper cannot be overstressed. A poorly aligned pilot hole can cause binding or drifting of the guide tube, with the resultant possibility of blade warpage or other damage.

Future Plans

Additional refinements are being made to the equipment, such as the installation of self-aligning blade shaft bearings and an improved roller chain lubrication system. These and other modifications will be made preparatory to the full-scale tests planned for the forthcoming construction season. Other systems of measuring rock mass movement to be used as backups to the multiple-position calipers are also under study.

Although the first phase of this program—the design, construction, and field testing of necessary equipment—is nearing completion, it is recognized that the more formidable problem lies in the interpretation of data. A number of unknowns remain to be answered in this respect, such as the effect of the cased caliper holes and the 7 ¾-in.-diameter pilot holes on the stress distribution around the saw cuts, the time-dependent behavior of the rock during the unloading and loading cycles, and the relationship of the *in situ* stress and deformation modulus to the pressures and deformations produced by the flatjacks. It is hoped that a better understanding of these unknowns will result from planned analytical and empirical studies.

The conclusions reached in these studies will be verified by large-scale laboratory tests of a stressed material of known modulus. These are the topics of present planning and, along with the results of full-scale field tests, will be the subject of a future report.

Acknowledgments

The equipment described and the data presented were obtained from research conducted under the Research, Development, Testing, and Evaluation Program (RDT & E) of the U. S. Army Corps of Engineers at the Missouri River Division Laboratory. The permission granted by the chief of engineers to publish this information is appreciated.

The authors also express their appreciation to K. S. Lane, Chief, Geology, Soils and Materials Branch, Corps of Engineers, Missouri River Division, for his encouragement and constructive criticism on this project. Special thanks are given to R. F. Anderson, J. Nielsen, and M. G. Ord, Missouri River Division Laboratory, for their very capable and skillful assistance in the design and fabrication of the apparatus.

S. J. Dixon[1]

Pressure Meter Testing of Soft Bedrock

REFERENCE: Dixon, S. J., "**Pressure Meter Testing of Soft Bedrock,**" *Determination of the* In Situ *Modulus of Deformation of Rock, ASTM STP 477,* American Society for Testing and Materials, 1970, pp. 126–136.

ABSTRACT: A description of the equipment and theory of an expansion pressure meter used directly in exploratory boreholes in rock is presented. The instrument shows promise in obtaining accurate and useful information for the stress-strain characteristics of soft bedrocks at economical cost. Test results of two of the first major engineering applications of the instrument in the United States are described.

KEY WORDS: pressure gages, pressure, expansion, modulus of elasticity, rock mechanics, rocks, stress analysis, shear properties, boreholes, foundations, compressive properties, testing

Engineers have long realized the desirability and importance of evaluating the engineering properties of soil and bedrock in their least-disturbed condition. There is a continuing need for methods which utilize practical and economic types of equipment for foundation testing. Development of new testing devices has not raised much interest in our universities, and practicing engineers seem to be content to rely on the same old standbys. In the field of soil mechanics, the most popular *in situ* methods are the standard penetration test (N value), the vane shear test, and the static cone penetrometer. Conventional laboratory strength and compressibility tests on undisturbed specimens are also used with varying degrees of success.

In rock mechanics the alternatives are expensive and troublesome, for it has been necessary to rely on plate bearing tests, flatjacks, extensometers, and other specially fabricated equipment. Also, it is normally necessary to make major exploratory excavations in which rock tests can be performed, and a project must be large to justify the expense. Laboratory tests have been only partially successful in indicating field conditions.

[1] Principal engineer, Converse, Davis and Associates, Pasadena, Cal. 91105.

It is not necessary to describe the above-mentioned methods as they are adequately considered in many technical papers. The important fact is that none of them were designed primarily to investigate the properties of soft rocks. There are few procedures available, except empirical practice, for investigating the qualities of soft rock. This material is too strong to undergo testing by blow count, vane shear, penetrometer, or to acquire undisturbed drive specimens, and is usually too weak to provide representative core samples. It is believed that a pressure meter developed by L. F. Menard[2] will help fill this gap.

The pressure meter is designed for direct use in exploratory boreholes and consists of a device that can be lowered into a borehole and expanded to contact and press against the borehall wall, giving the stress-strain characteristics of the material. Elastic and plastic properties can be determined from the recorded data.

The idea for the pressure meter was first conceived many years ago, but it was not until the 1950's that the technical and theoretical concepts of the present instrument were developed. The device was first used in an experimental form in the early 1960's. Since the pressure meter became commercially available in 1966 to engineers in the United States, it has shown promise in investigating soft bedrocks. Its main advantages are those of economics and the ability to record *in situ* properties of materials that are weathered, fractured, difficult to sample, or just generally weak but intact. The materials tested to date have consisted primarily of shales, sandstones, and mudstones.

The application of the instrument is as yet in the experimental stage in the United States. There is still little published data relating the results of pressure meter testing to prototype performance or comparing pressure meter tests with other testing results. It is important that this lack of knowledge be recognized and that engineers report their results wherever possible. The projects described in this paper were two of the first major engineering applications of the pressure meter equipment in the United States.

Description of Equipment

The pressure meter equipment, shown schematically in Fig. 1, consists of a combination volumeter-manometer connected to a cylindrical borehole expansion device, or probe. The probe consists of a steel tube surrounded by two flexible rubber membranes. The interior membrane forms the measuring cell, and the exterior membrane provides guard cells at the two ends of the probe. The guard cell is activated by gas pressure and is

[2] Menard, L. F., "An Apparatus for Measuring the Strength of Soils in Place," MSCE thesis, University of Illinois, Urbana, Ill. 61801.

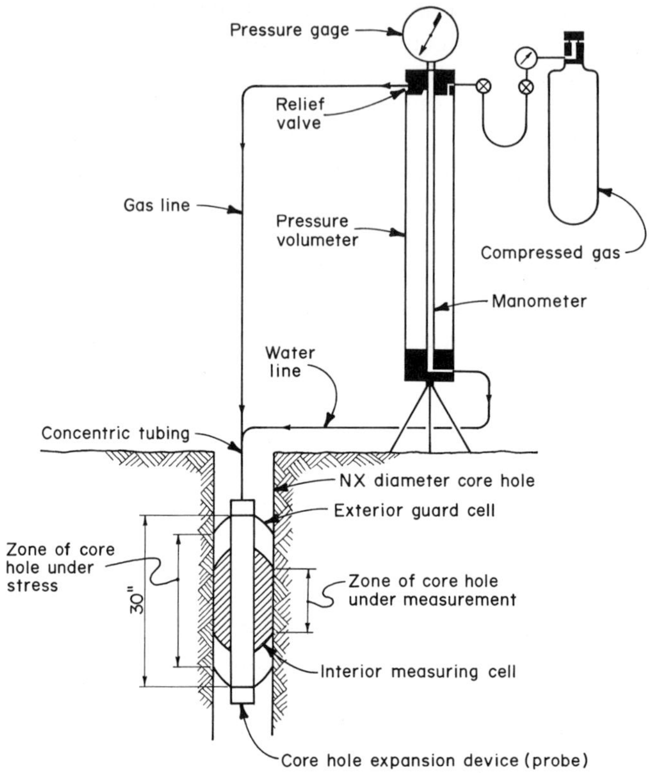

FIG. 1—*Schematic of pressure meter equipment.*

used to reduce end effects on the measuring cell to provide an essentially two-dimensional, cylindrical stress condition. The measuring cell is pressurized with water and kept at a slightly higher pressure than the guard cell to insure that it is always pressing against the borehole wall. Two concentric tubes connect the volumeter to the probe. An adapter connects the probe to a standard drill rod for lowering and raising the probe within the borehole.

After lowering the probe to the desired depth, pressure is applied to the borehole wall by inflating the rubber membranes. In addition to providing a means for applying increments of pressure to the measuring cell, the volumeter-manometer measures the radial expansion under each pressure increment by measurement of water flow into the probe.

Testing is normally performed during drilling. The simplicity of the equipment allows a great number of tests to be performed swiftly and economically without significant hindrance to the drilling operation. Testing does not interfere with the recovery of core samples, and by sample examination locations for testing can be preselected.

The pressure meter can be used inside EX, AX, and NX boreholes (1 7/16, 1 27/32, and 2 15/16 in. in diameter, respectively) to a significant depth. It is normally used in larger NX drill holes. It is adaptable for use in soil and soft rocks and other materials which have an *in situ* modulus of elasticity (deformation) of less than 1,000,000 to 1,500,000 psi, although values up to 3,000,000 psi can be obtained with a lesser degree of accuracy.

Calculation of Test Results

In a pressure meter test, a result of which is shown in Fig. 2, five basic quantities can be determined:

1. "Initial" pressure P_0—the beginning of the elastic stress range, that is, the seating pressure.
2. "Creep" pressure P_f—the end of the elastic stress range.
3. "Limit" pressure P_1—the failure pressure.
4. Compression modulus E—derived from the slope of the compression curve (between P_0 and P_f).
5. Rebound modulus E_r—derived from the slope of the rebound curve.

The P_0, P_f, *and* P_1 values are obtained directly from the stress versus deformation plot of volume versus corrected pressure. The modulus values are interpreted from the slope of the compression and rebound curves, which will be described below. The modulus values are calculated from

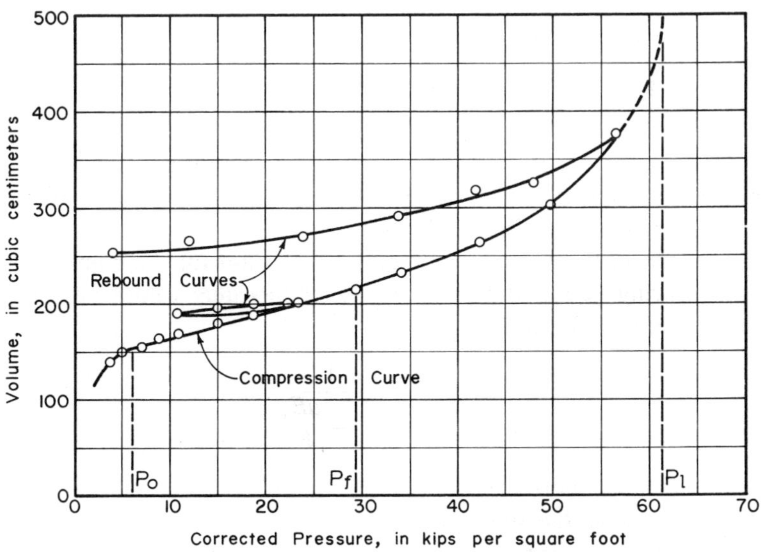

FIG. 2—*Typical pressure meter test results.*

the elastic equation of an infinitely thick cylinder subjected to internal pressure. The basic equation, as given in standard textbooks is

$$\delta = \frac{rp}{E}(1 + \mu) \qquad (1)$$

where:

δ = radial deflection,
r = radius of hole,
p = applied pressure (stress),
μ = Poisson's ratio (usually taken between 0.3 and 0.4), and
E = modulus.

It can be shown that the above equation, in terms of the measuring equipment, reduces to

$$E_1 = K\frac{dp}{dv} \qquad (2)$$

where:

E_1 = modulus,
dp = change in corrected pressure,
dv = volume change of probe, and
K = constant which includes the diameter of boring, the size of probe, and Poisson's ratio.

The corrected pressure is found as follows:

$$P = P_g - P_i + P_w$$

where:

P = corrected pressure,
P_g = gage pressure at ground surface,
P_i = inertia, and
P_w = net piezometric water head between the probe and gage.

The inertia is the pressure required to inflate the probe and varies with the inflated volume of the rubber membranes. It is determined by a calibration test on each probe assembly.

A correction to the E_1 value is needed to account for the compressibility of the measuring equipment; thus, the final modulus values are obtained from the following formula:

$$E = E_1 \frac{1}{1 - \frac{aE_1}{K}} \qquad (3)$$

where now E is the corrected modulus, a is the coefficient of (instrument) compressibility, and the other terms are as above.

The coefficient of compressibility, a, is generally independent of volume and is determined by a pressure-volume test in a rigid jacket. For most soft rock, the material is weak enough to compute accurate modulus values with one value of a for probes of equal dimension and constant length of connecting tubing. The total volume change is composed of

1. The actual borehole expansion, including the compression necessary to seat the probe against the borehole wall.
2. The compression of the probe membrane.
3. The measuring fluid compression.
4. The expansion of the connecting tube.

The last three items are found collectively by determining the coefficient of compressibility. The expansion of the connecting tube is reduced to a small quantity by the use of the concentric tubing, since the pressure in the interior measuring tube is balanced by the gas pressure in the exterior tube.

In the calculation of the modulus value, it should be noted that the stress condition in the borehole walls will cause radial cracking when the testing pressure exceeds the tensile strength of the material. This condition will lead, using the formulas described, to an underestimation of the modulus values. It is also important to determine modulus values, if possible, for stress ranges at least as high as those expected in design analysis, for tranformation of test results to higher pressures is not possible.

Castaic Surge Chamber Testing

An important part of the West Branch of the California Aqueduct is the Angeles Tunnel and 1200-MW Castaic Power Plant located about 50 miles north of Los Angeles. The Castaic Surge Chamber, a necessary feature of the power plant, is a differential-type chamber with an inside diameter of 120 ft and a total height of 390 ft. The structure extends approximately 235 ft below the ground surface and is connected to a power tunnel by a 28-ft-diameter riser shaft. All excavation was in interbedded sandstone and shale of the Castaic formation. Construction of the tunnel and surge chamber began early in 1967.

Due to the unprecedented size of the surge chamber, it was decided during design studies to use several methods to evaluate the engineering qualities of the foundation material. Three independent methods were utilized:

1. The pressure meter first developed by L. F. Menard.
2. Laboratory testing of rock core specimens by static and dynamic methods.
3. Geophysical seismic studies.

The pressure meter tests were made with the aid of conventional drilling equipment in a 400-ft NX core hole. After drilling each run of approximately 20 ft and examining the recovered core, the pressure meter probe was positioned for testing at selected depths. A total of 63 tests was made in a depth range of 75 to 400 ft.

The pressure meter tests were performed by applying pressure increments of 15 to 75 psi, increasing to a maximum of 600 psi in the hardest zones and 300 psi in the weakest zones, and then taking volume change readings 15, 30, and 60 s after each pressure application. The rebound test was performed in pressure decrements of 70 psi, with readings taken at 15 s.

The modulus values were computed for the initial compression and rebound phases of the test. In general, the compression and rebound moduli were determined from stress ranges of about 100 to 500 psi and 500 to 200 psi, respectively. A summary of the final test results, including laboratory strength and compression moduli from unconfined tests and density data, is shown in Fig. 3.

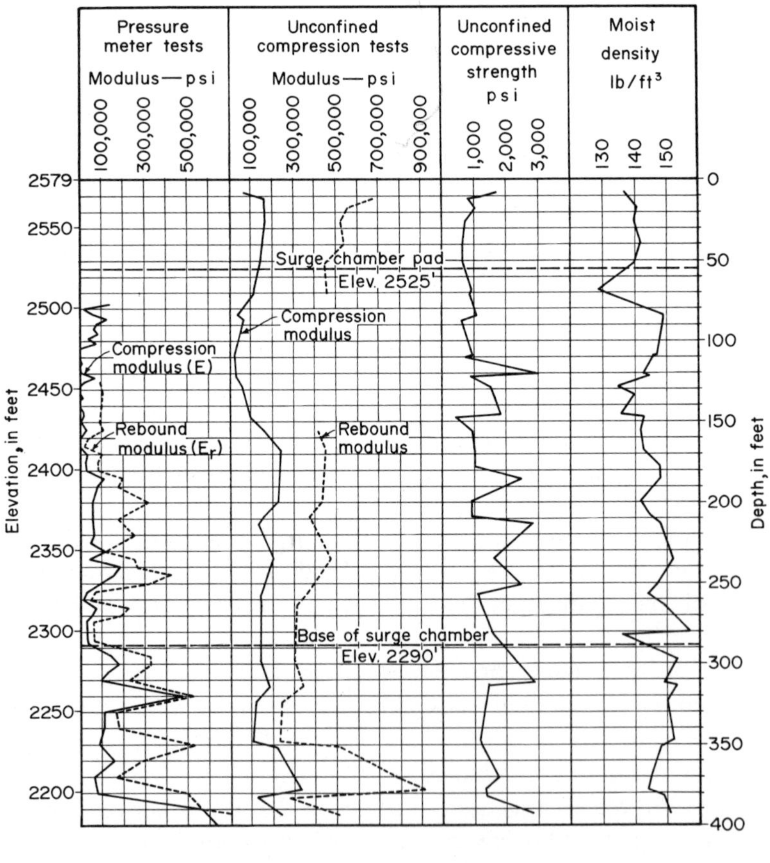

FIG. 3—*Summary of Castaic Surge Chamber test results.*

The most noticeable trend in the modulus values was the variation in material properties with depth. Both the laboratory compression and pressure meter tests indicated low-quality material to a depth of about Elevation 2400. From Elevation 2400 to 2230, the quality improved and was reasonably consistent; below Elevation 2230, the material performed as a fresh, relatively unfractured rock. The pressure meter results showed consistently lower modulus values in the sandstone layers, and averages of both the field and the laboratory tests indicated that the modulus values of the sandstone were on the order of one half of those of the siltstone. The average compressive strengths and densities in the sandstone were also lower than the siltstone. The lower values in the sandstone are most clearly seen in pressure meter tests in the zones of Elevation 2475 to 2400, Elevation 2330 to 2315, Elevation 2310 to 2290, and Elevation 2250 to 2200.

As in all cases of laboratory testing of rock samples, the specimens selected were necessarily of better average quality than the overall material. In the sandstone and siltstone materials this was particularly true, as the core samples rapidly deteriorated on exposure to the atmosphere, and only the better samples were available for testing purposes. This fact, plus the influence of jointing and fracturing on the deformation properties, demonstrated the importance of the field tests. The higher values of the pressure meter results approached those of the laboratory tests, but the lower pressure meter test results were considered more indicative of the average *in situ* properties.

A comparison was made of the ratio of the rebound to the compression modulus values for both the field and laboratory tests, for it provided a qualitative measure of the degree of fracturing of the rock. The ratio of E_r to E for the field tests varied from a minimum of about 1.5 to as high as 10. The results did indicate and confirm visual interpretation of zones of moderately to heavily fractured rock.

One of the most important design parameters for the Castaic Surge Chamber was the modulus of deformation, since foundation movements were of prime concern. The pressure meter test results served as the primary basis for the selection of deformation properties. Foundation analysis of the surge chamber and its excavation included the determination of allowable bearing, settlement under vertical loading and unloading, and lateral resistance and movement under the action of both hydrostatic and dynamic lateral forces. The pressure meter tests were also of value in assessing construction support requirements for the surge chamber excavation.

Stokes Canyon Tunnel Testing

An auxiliary link of California's current water diversion system is the proposed Stokes Canyon Dam and Inlet-Outlet Tunnel located west of Los Angeles. In the course of previous investigations in the project area, it was found that the proposed ½-mile tunnel would be located through marine sediments consisting of interbedded sandstones and clayey siltstones. Supports that had failed in two exploratory drifts indicated potential problems; this made it desirable to estimate the probable loads for design of temporary steel supports in the proposed 9-ft-diameter tunnel with the pressure meter method of Menard.

During 1967 pressure meter tests were performed in a 166-ft-deep NX core hole. The ground water table was about 50 ft above tunnel level. The pattern of drilling and testing was to drill 25 ft with a roller bit, take a 5-ft NX core, and perform a pressure meter test in the cored interval. This process was continued until the hole reached a point 10 ft above the tunnel crown, whereupon a continuous core was taken and pressure meter tests were made at approximately 5-ft intervals to a point 5 ft below the tunnel invert. Ten pressure meter tests were performed, whose results are presented in Table 1.

TABLE 1—*Pressure meter test results for Stokes Canyon Tunnel boring No. C-11.*

Depth, ft	Initial Pressure P_0, ksf	Creep Pressure P_f, ksf	Limit Pressure P_1, ksf	Compression Modulus E, psi	Rebound Modulus E_r, psi
32.5	5	32	70	14 800	37 400
57.5	5	15	30	5 150	11 600
87.0	6	38	90	12 100	38 100
115.0	2	11	50	9 800	...
137.5	6	30	70	9 400	19 300
145.0	10	22	...	15 000	20 900
149.0	5	13	40	8 300	...
153.0	5	20	50	9 800	14 600
158.0	8	22	50	17 800	...
163.0	6	21	45	10 900	14 000

NOTE—All material was predominantly clayey siltstone, with moist densities varying from 130 to 140 lb/ft^3.

The pressure meter tests provided an estimate of the *in situ* lateral pressure, strength, and deformation properties of the material. Initial pressure P_0 near the tunnel elevation, as measured with the pressure meter, ranged from 5 to 10 ksf. Creep pressure P_f ranged between 11 and 38 ksf. The compression modulus values varied from about 5000 to 18,000 psi, and the rebound modulus from about 12,000 to 40,000 psi. The ratio of the rebound modulus to the compression modulus ranged between 1.3 and 3.1, indicating a generally intact, soft bedrock.

Laboratory tests consisting of clay mineralogy, consolidation, swell, and strength tests provided additional information on the material properties.

The results of the pressure meter tests were indicative of soft bedrock. No significant trend in the modulus values with depth was evident, except that the ratio of the rebound modulus to the compression modulus was larger at the shallow depths—confirming a visual inspection of the core that suggested the extent of fracturing decreased with depth.

The pressure meter tests provided an estimate of the tunnel support loads and their increase with time. Although a discussion of the analysis is beyond the scope of this paper, the results indicated that temporary supports should be designed for a crown pressure of 5000 psf, a side pressure of 2500 psf, and an invert pressure of 3000 psf, and that support loads would increase to a maximum in two to three weeks. The field and laboratory testing also enabled a determination of the squeezing and swelling potential of the bedrock, and it was concluded that the materials exhibited predominantly swelling properties.

In comparing the results of the investigation with the practices in use in similar materials, the tunnel design loads were moderately severe for the size of tunnel being considered, but not as severe as some examples of tunnels in swelling ground.

Conclusions

In both the Castaic and Stokes Canyon projects the results of field testing with the pressure meter were given credence over other test methods although greater use of the equipment is required to gain more confidence in the testing results and application. Laboratory tests were performed to provide useful data for substantiating the ability of the equipment to determine the engineering characteristics of soft bedrock.

The cost of obtaining the information by use of the pressure meter was quite low. On the Castaic Surge Chamber the total cost per test, excluding drilling, was less than $50. The minimal cost, together with the importance and desirability of measuring the engineering properties of soft bedrock *in situ* by many tests, is a clear advantage of the equipment. Thus, the pressure meter should be used as a testing tool wherever possible so that a greater amount of experience can be collected and made available.

Acknowledgments

Enginering design for both the Castaic Surge Chamber and the Stokes Canyon Tunnel was performed by the engineering staff of the Department of Water and Power, City of Los Angeles. Converse, Davis and Associates served as engineering and geologic consultants to the City. Shea, Kaiser, Lockheed, and Heely are the general contractors; and the State of California Department of Water Resources administers the construction contract for the Castaic Surge Chamber. Appreciation is expressed to B. E.

Hartmann, who was helpful in establishing the initial procedures for the use of the pressure meter.

Bibliography

[1] Dixon, S. J. and Jones, W. V., *Civil Engineering—ASCE,* Vol. 38, No. 10, Oct. 1968, pp. 34–36.

[2] Gibson, R. E. and Anderson, W. F., *Civil Engineering and Public Works Review,* CEWRA, May 1961, pp. 3–6.

[3] Higgins, C. M., "Pressure Meter Correlation Study," *Highway Research Record No. 284,* HIRRA, Laboratory and Field Soil Tests, 1969, pp. 51–62.

[4] Ladanyi, B., "Evaluation of Pressure Meter Tests in Granular Soils," *Proceedings,* Second Panamerican Conference on Soil Mechanics and Foundation Engineering, Brazil, 1963, Vol. 1, pp. 1–20.

[5] Menard, L. F., "Calculation of the Bearing Capacity of Foundations Based on the Results of Pressure Meter Tests," *Sols - Soils,* Paris, Vol. 2, No. 5, June 1963, pp. 9–32.

[6] Menard, L. F., "Calculation of the Bearing Capacity of Foundations Based on the Results of Pressure Meter Tests," *Sols - Soils,* Paris, Vol. 2, No. 6, September 1963, pp. 9–31.

[7] Menard, L. F., "The Application of the Pressure Meter for Investigation of Rock Masses," *Colloquium of the International Society for Rock Mechanics,* Salzberg, Austria, 1965.

[8] Menard, L. F., "Rules for the Calculation of Bearing Capacity and Foundation Settlement Based on Pressure Meter Tests," *Proceedings,* Sixth International Conference on Soil Mechanics and Foundation Engineering, Vol. 2, Montreal, 1965, pp. 295–299.

[9] Menard, L. F., "The Geocel Pressure Meter. Interpretation of a Pressure Meter Test," unpublished; available from Geocel, Inc., Golden, Col. 80401.

[10] Mori, H. and Tajima, S., *Soil and Foundation,* SOLFA, Tokyo, Vol. 4, No. 2, Feb. 1964, pp. 34–44.

A. J. Hendron, Jr.,[1] *G. Mesri,*[1] *J. C. Gamble,*[1] *and G. Way*[2]

Compressibility Characteristics of Shales Measured by Laboratory and *In Situ* Tests

REFERENCE: Hendron, Jr., A. J., Mesri, G., Gamble, J. C. and Way, G., **"Compressibility Characteristics of Shales Measured by Laboratory and *In Situ* Tests,"** *Determination of the* In Situ *Modulus of Deformation of Rock, ASTM STP 477,* American Society for Testing and Materials, 1970, pp. 137–153.

ABSTRACT: A comparison of laboratory tests and *in situ* pressure meter tests shows that the pressure meter can be used to infer the compressibility properties of shales *in situ*. The moduli values computed directly from the pressure meter tests, however, overestimate the "drained" modulus which must be used for design, and it is suggested that the appropriate deformation modulus for use in settlement calculations be taken as about one third of the moduli measured with the pressure meter. Laboratory and *in situ* tests indicate that both the moduli and strength can be correlated with initial water content for a given shale. These correlations can be used to estimate the moduli at locations on the basis of a water content profile in the absence of laboratory or pressure meter tests. Laboratory tests also indicate that the ratio of the deformation modulus to the compressive strength is a constant for a given shale; thus, the compressive strength can also be used as an index property to estimate the compressibility of a shale.

KEY WORDS: shale, modulus of elasticity, pressure gages, laboratories, compressibility, tests, moisture content, compressive strength, foundations

Soft shales and underclays of the Pennsylvanian period are frequently encountered in the foundations of bridge piers, dams, and lock structures in the Ohio and Mississippi River valleys. In order to assess the long-term differential settlements, the compressibility characteristics of these materials are usually determined by laboratory or *in situ* tests or empirical

[1] Associate professor, assistant professor, and research assistant, respectively, Department of Civil Engineering, University of Illinois, Urbana, Ill. 61801.

[2] Chief soils and materials engineer, De Leuw, Cather & Co., Engineers, Chicago, Ill. 60601.

extrapolation of settlement measurements from structures founded on similar materials.

Laboratory tests are probably the most common means of estimating the compressibility of "clay shales"; but the value of the tests is greatly suspect because of sample disturbance. Although field load tests are sometimes employed to circumvent the problem of sample disturbance, sites for many of the structures mentioned are located underwater and field load tests are not possible. The pressure meter is a relatively new tool which may be of significant value in determining the compressibility of these materials by testing the shale *in situ*. The pressure meter can be lowered down an NX core hole to obtain a continuous record with depth, even at locations where it is not possible to recover samples.

In this paper, compressibility data are presented from laboratory and pressure meter tests conducted on shales from the Mississippi River crossing of Interstate 280 near Rock Island, Ill., where compressibility properties were needed to estimate the possible differential settlement between piers of a continuous, multiple-span highway bridge. Deformation moduli measured by the pressure meter and laboratory tests are compared, and correlations are also given between measured deformation moduli and index properties such as the initial water content and undrained shear strength of the shale. These correlations can be used to estimate deformation moduli at locations where only the index properties are available.

Preliminary Considerations

Geologic Description and Index Properties of Shales Tested

The bridge for Interstate 280 crosses the Mississippi River about one mile downstream from the Rock River juncture. Alluvial deposits from the Mississippi River ranging from 4 to 40 ft in depth overlie Pennsylvanian period sedimentary rocks with thicknesses of 40 to 60 ft, which in turn overlie Middle Devonian period limestone, as shown in Fig. 1. The bridge piers are founded on the Pennsylvanian rocks (Caseyville Formation), which consist primarily of gray silty shale and dark-gray to black fissile shale, with lesser amounts of coal, underclay, sandy shale, siltstone, sandstone, and conglomerate. A facies change in the shales occurs in about the middle of the river, with gray shales on the Iowa side and black shales on Illinois side. Generally, the upper beds dip about 6 deg in the upstream direction.

The gray shales tend to be somewhat massive except for abundant plant material found along many bedding planes. Grain size analyses given in Fig. 2 show compositions of 3 to 6 percent sand, 58 to 73 percent silt, and 23 to 39 percent clay ($<$ 0.002 mm). Typical Atterberg limits are liquid limit, 37 percent, and plastic limit, 22 percent.

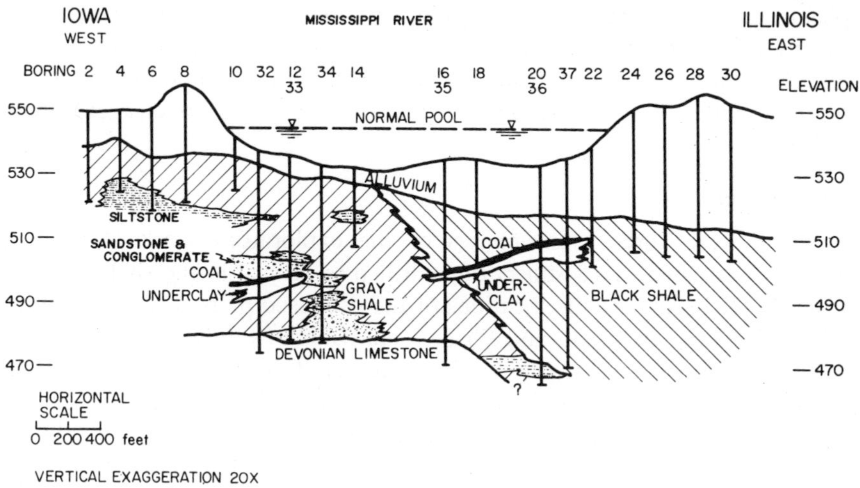

FIG. 1—*Generalized geologic profile and soil borings.*

Most of the underclays have grain size distributions and Atterberg limits similar to the gray shales. The dark-gray to black shale is fissile and generally blocky. It contains 13 to 21 percent sand, 37 to 47 percent silt, and 40 to 47 percent clay, except for a soft upper layer which contains only a trace of sand. Typical limit values are liquid limit, 45 percent, and

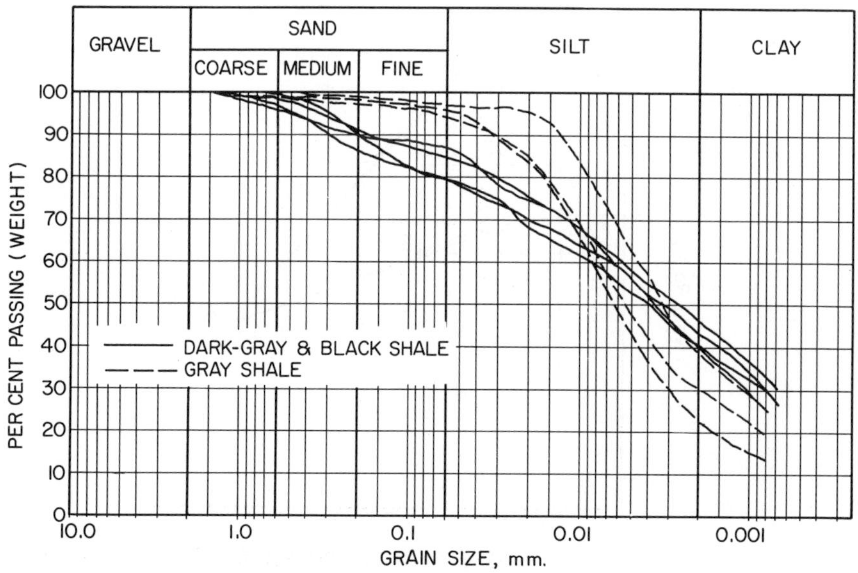

FIG. 2—*Grain size curves for gray shale and dark-gray to black shale.*

plastic limit, 26 percent. Clay mineralogy analyses reveal the clays to be mainly illite and chlorite with lesser amounts of kaolinite and mixed-layer clay minerals (Table 1). Both black and gray shales and the underclay have a tendency to crack upon exposure to air and to disaggregate in water or slake.

Exploration

There were three main phases of the exploratory work at the site: Phase 1 consisted of 18 soil borings drilled to refusal in the shale; Phase 2 had 32 NX borings at the locations of the proposed piers, which penetrated 10 to 20 ft into the rock; and Phase 3 consisted of 6 NX borings wherever soft zones or low recovery were encountered in the previous borings. These were cored about 50 ft deep into rock, and pressure meter tests were run at approximately 5-ft intervals after each core run. Six 5½-in.-diameter core borings were then drilled 10 to 15 ft away from the NX to provide core for laboratory testing. All laboratory tests reported in this study were conducted on the NX and 5½-in. cores.

Mechanical Properties Required for Settlement Prediction

The long-term settlement of a bridge pier is predicted by numerically integrating the vertical strains in elements of shale below the pier base,

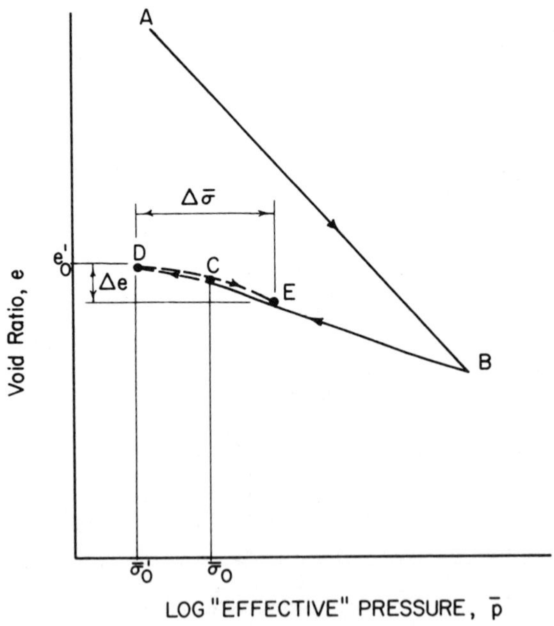

FIG. 3—*Schematic representation of void ratio versus* log *"effective" pressure.*

TABLE 1—*Clay mineral analyses.*[a]

Boring	Depth, ft	Rock Species	Clay Minerals, Parts in 10 of Diffraction Effects			
			Illite	Mixed-Layer Clay Minerals	Chlorite	Kaolinite
32A	27.8 to 28.5	underclay	4 to 5	2	2	1
33	28.3 to 28.8	gray clayey shale	5	2	total of both is 3	
34A	28.1 to 29.0	gray shale	6	1 to 2	. . .	2 to 3
35	56.1 to 56.4	gray shale	6	1	2 to 3	1
35A	40.3 to 41.3	black shale	4	1	3	2
	47.5 to 48.0	underclay	4 to 5	2 to 3	2	1
36A	50.4 to 51.4	black shale	4	3	2	1
37A	36.1 to 37.2	black clay shale	4	2	2	1
	39.7 to 40.2	underclay	5	2 to 3	. . .	2 to 3
	52.7 to 53.7	black shale	4 to 5	1	total of both is 3 to 4	

[a]Performed courtesy of Illinois State Geological Survey by W. A. White.

which result from the changes in effective stress imposed by the dead weight of the pier and the bridge reaction. Each element of shale below the pier foundation is material which has been heavily preloaded by possibly thousands of feet of sediments that have been subsequently eroded. This simplification of the stress history is shown on a schematic plot of void ratio versus log "effective" stress in Fig. 3, where the loading portion is shown along the path *AB* and the erosion, or rebound, portion of the curve is shown by the curve *BC* which terminates at the present "effective" stress $\bar{\sigma}_0$ to which the shale is subjected. The excavation for the pier further reduces the effective stress on the element to $\bar{\sigma}_0'$ to point *D*; the construction of the pier and the live load of the pier recompress the shale to point *E*. Thus all the deformations occur on a recompression curve at loads much below the maximum preload.

The laboratory tests used to estimate the deformation moduli appropriate for design should therefore be conducted at initial "effective" confining pressures near the value of $\bar{\sigma}_0'$ or $\bar{\sigma}_0$. The moduli of interest for design is the ratio of the change in "effective" stress $\Delta\bar{\sigma}$ (Fig. 3) to the increment of strain which occurs under the stress $\Delta\bar{\sigma}$ from the time the load increment is applied until 100 percent primary consolidation, as defined by Terzaghi and Peck [*1*].[3]

In Situ and Laboratory Tests

The stress-strain properties of the shales described above were determined by one-dimensional consolidation tests, load-controlled triaxial tests, strain-controlled triaxial tests, and *in situ* pressure meter tests.

The type of loading and boundary conditions for each of these tests is shown schematically in Fig. 4. It should be noted that "drained" stress-strain characteristics are obtained from the one-dimensional and stress-controlled triaxial tests. "Undrained" moduli are measured in the strain-controlled triaxial tests, the stress-controlled triaxial tests, and pressure meter tests.

One-Dimensional Consolidation Tests

The one-dimensional consolidation tests were performed in a conventional manner [*2*] on undisturbed specimens of shale trimmed into 2½-in.-diameter Wykeham Furnace consolidation rings ¾ in. high. The void ratio versus log pressure curves (*e*-log *p*) for two of these tests are shown in Fig. 5. The same data plotted in the form of stress-strain diagrams are shown in Fig. 6. The compression indexes (C_c), taken as the slope of the steep portion of the *e*-log *p* curves, range from 0.06 to 0.07. The very

[3] Italic numbers in brackets refer to the list of references at the end of this paper.

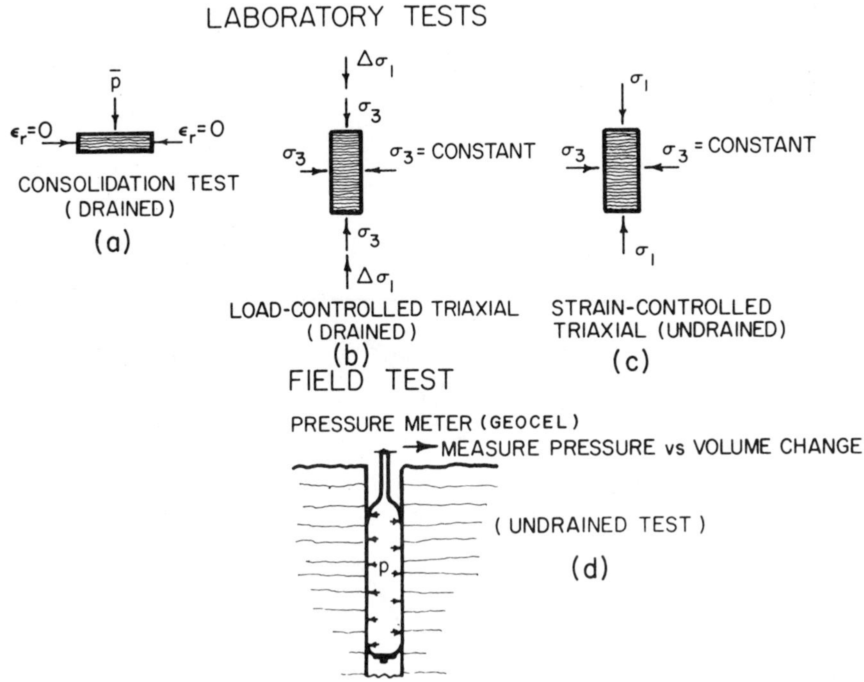

FIG. 4—*Schematic representation of loading and boundary condition in laboratory and* in situ *tests.*

low values of the compression indexes and the close proximity of the rebound and loading portion of the *e*-log *p* curves, which show nearly elastic behavior, indicate that the shales had been preloaded to much higher loads than they were subjected to in these tests. The one-dimensional stress-strain curve for each test (Fig. 6) is a curve which increases in stiffness as the stress increases and is thus concave to the stress axis. On each stress-strain curve the secant modulus is shown which best represents the stiffness of the shale for the pressures applied by the pier loads.

Stress-Controlled Triaxial Tests

The constant load, triaxial compression tests were performed on undisturbed shale specimens 1⅞ in. in diameter and 3⅓ in. high. The specimens were set up in the Wykeham Farrance triaxial cells and allowed to come to equilibrium under an all-around confining pressure. The first group of specimens was subjected to a confining pressure equivalent to the present overburden pressure in the field. The shale specimens swelled under this confining pressure. The confining pressures of a second group of specimens were increased until no swelling occurred. After the specimens

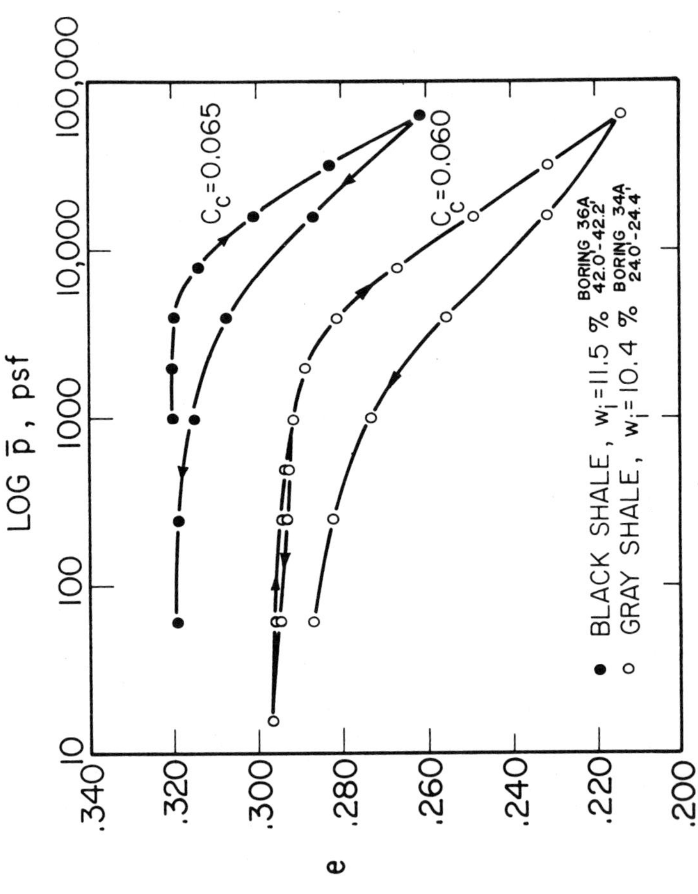

FIG. 5—*Void ratio versus* log *pressure relationship from one-dimensional consolidation test.*

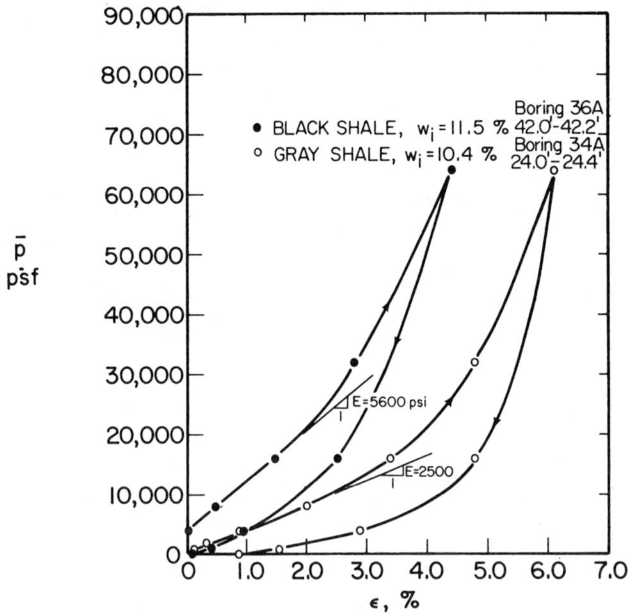

FIG. 6—*Stress-strain relationship from one-dimensional consolidation test.*

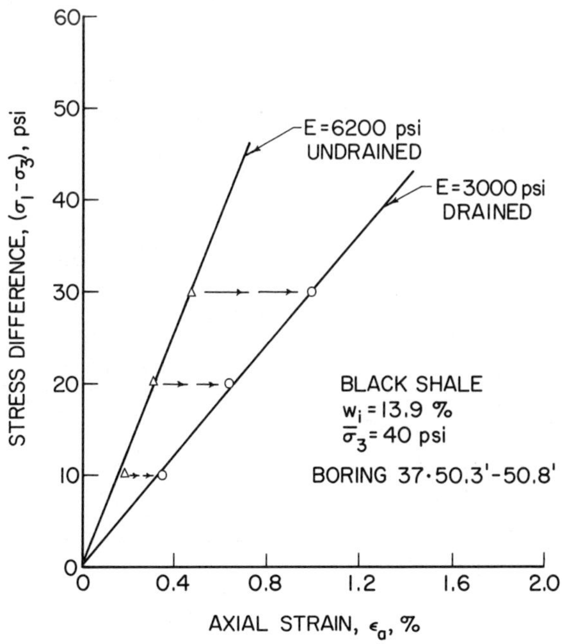

FIG. 7—*Stress-strain relationship from stress-controlled triaxial compression test.*

had reached equilibrium under the confining pressure, the drainage valve was closed and the first increment of axial load was applied by placing weights on a platform which hung on the piston of the triaxial cell. An axial deformation reading was taken under undrained conditions and used to calculate undrained moduli. The drainage valve was opened, and axial deformation and volume change measurements were made and used to calculate drained moduli. The next increment of the axial load was applied, and the procedure repeated. Typical drained and undrained stress-strain curves obtained from a stress-controlled triaxial test are shown in Fig. 7 for a specimen of black shale consolidated to 40-psi confining pressure to prevent swelling of the specimen during the consolidation phase of the test.

Strain-Controlled Triaxial Tests

Unconsolidated-undrained triaxial compression tests were conducted on 5½ by 8-in. specimens of shale to determine the undrained moduli as well as the undrained shear strength. The specimens were tested at a confining pressure of 10 psi and loaded to failure by applying a stress differ-

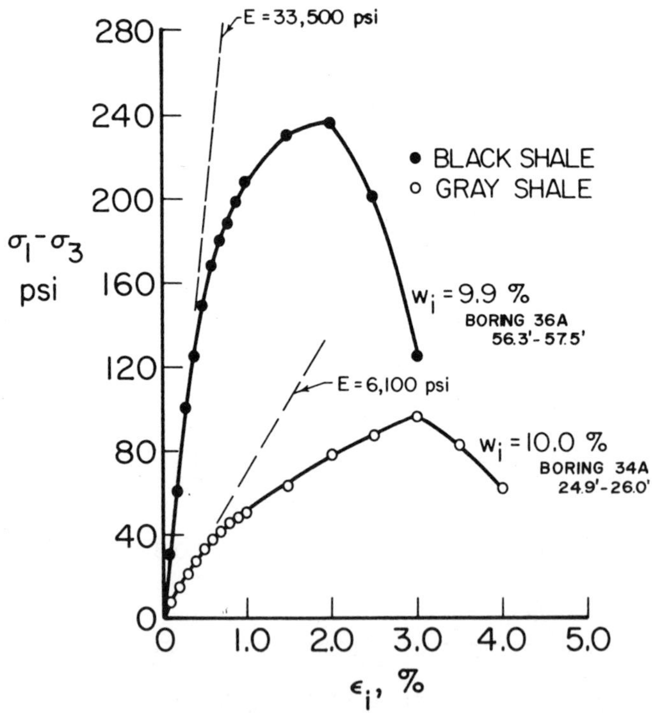

FIG. 8—*Stress-strain relationship from strain-controlled triaxial compression test.*

ence at a deformation rate of 0.04 in./min. Typical stress-strain curves for specimens of a black shale and a gray shale are shown in Fig. 8. The initial slope of these curves is the undrained initial tangent modulus, E_i, and the maximum value of $(\sigma_1 - \sigma_3)$ is taken as the undrained strength.

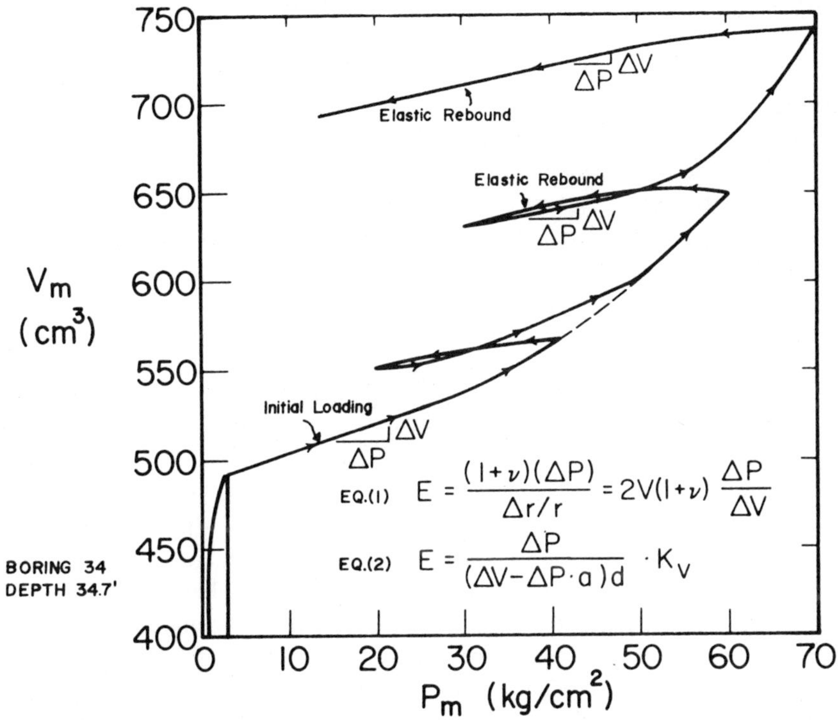

FIG. 9—*Volume change versus pressure relationship from pressure meter test.*

Pressure Meter Tests

Pressure meter tests were conducted at 5-ft intervals in six NX holes. A typical pressure meter test result is shown in Fig. 9, which gives the volume change of the cell versus pressure applied to the cell. The deformation modulus of the shale on initial loading is related to the initial slope $\frac{\Delta V}{\Delta P}$ shown in Fig. 9. For a pressure meter of volume V the modulus of the material surrounding a drill hole of radius r is given by

$$E = \frac{(1 + \nu)\,(\Delta P)}{\Delta r/r} = 2V\,(1 + \nu)\frac{\Delta P}{\Delta V} \quad \text{......................(1)}$$

where ν is Poisson's ratio of the medium.

Equation 1 is derived for the ideal condition that the volume change shown in Fig. 9 actually represents a volume change of the length of drill hole compressed by the pressure meter. These apparent volume changes must be corrected for the compressibility of the tubing, the compressibility of the fluid, and the compressibility of the pressure meter. Applying these corrections to Eq 1 yields an equation of the form:

$$E = \frac{\Delta P}{(\Delta V - \Delta P\, a)\, d} \times K_v \quad \text{.........................(2)}$$

where:

ΔV = the variation in measured volume,
ΔP = the variation in measured pressure,
a = the entire system compressibility factor,
d = the correction factor for system volume error, and
K_v = the compression coefficient. (This constant is the initial volume of the cell plus V_m shown on Fig. 9 times $(1 + \nu)$.)

The values of the moduli reported from the pressure meter tests were calculated by means of Eq 2 using constants appropriate for the pressure meter apparatus used. More complete descriptions of the pressure meter and the derivation of Eq 2 are given by Menard [*3*] and Dixon and Jones [*4*].

The value of the deformation modulus of interest in this study is the stiffness on initial loading, not the rebound moduli shown in Fig. 9. It should be noted that the pressure meter test is conducted by applying the pressure in increments and then reading the volume change at 15, 30, and 60 s after a constant pressure is reached. Thus, the moduli determined from the pressure meter test are undrained values of the modulus. These values should probably be higher than the undrained values determined from the laboratory, because the stress difference on the laboratory specimens is applied perpendicular to the laminations in the shale, whereas in the pressure meter test the pressure is applied parallel to the laminations. In addition, the values for the laboratory undrained specimens are lower than the pressure meter values because of specimen disturbance.

Discussion of Test Results

The purpose of this investigation was to determine the deformation moduli of the shales so that the settlement of many bridge piers could be calculated. The number of tests conducted on this study was obviously not adequate to completely describe the stress-strain characteristics of the shales below each pier. But from the outset of the investigation, tests were conducted on both soft and hard specimens of shale in an attempt to get a correlation of the measured moduli and a readily determined index property such as the initial water content.

Previous studies of the relationship of deformation modulus with strength

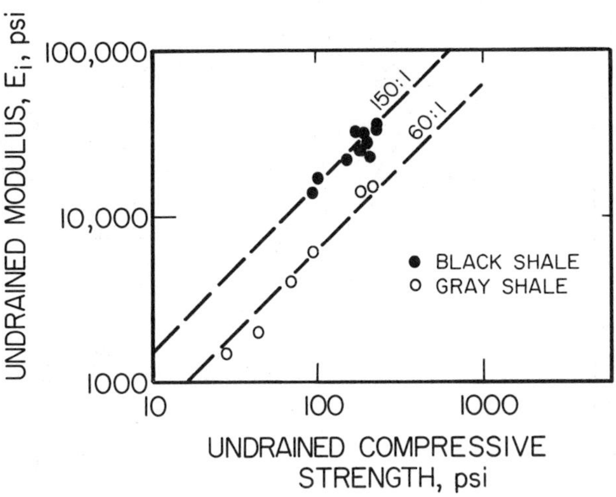

FIG. 10—*Undrained modulus versus undrained compressive strength for gray and black shales.*

of soils and rock by Wilson and Dietrich [*5*] and Deere and Miller [*6*] have shown that shear strength and deformation moduli are related. A plot of the undrained Young's modulus versus the undrained compressive strength is shown in Fig. 10 for the shales tested in this investigation. The relationship shown indicates that the ratio of undrained Young's modulus to the undrained compressive strength is 150:1 and 60:1 for the black shale and gray shale, respectively. Previous studies on soil have also shown that the undrained shear strength of a saturated soil is related to the water content or void ratio before shearing. Since the relationships shown in Fig. 10 substantiate the finding that deformation moduli and strength are related by a constant, then deformation moduli should also be a function of initial water content.

Figure 11 shows the relationship of log $(\sigma_1 - \sigma_3)_f$ versus water content for undrained tests on gray shale. It is observed that the relationship is a straight line on the semilog plot similar to the virgin portion of an *e*-log *p* curve for a consolidation test. A plot of log E_i versus water content shown in Fig. 11 for undrained tests on the gray shale plot is a straight line parallel to the log strength versus water content relationship. A plot of log drained moduli versus water content for two triaxial tests and two consolidation tests on the gray shale also gives a straight line parallel to the other relationships shown in the figure. It should be noted that the drained modulus is roughly one half of the undrained modulus for a specimen at a given initial water content. The drained modulus will determine the total settlement of the pier, and the information contained in Fig. 11 will enable the drained modulus to be estimated at a given location by a

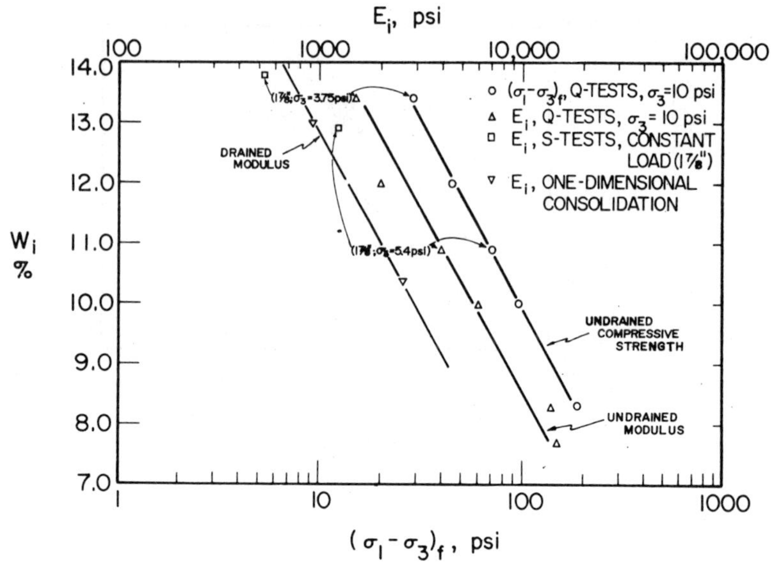

FIG. 11—*Modulus and compressive strength versus water content for gray shale.*

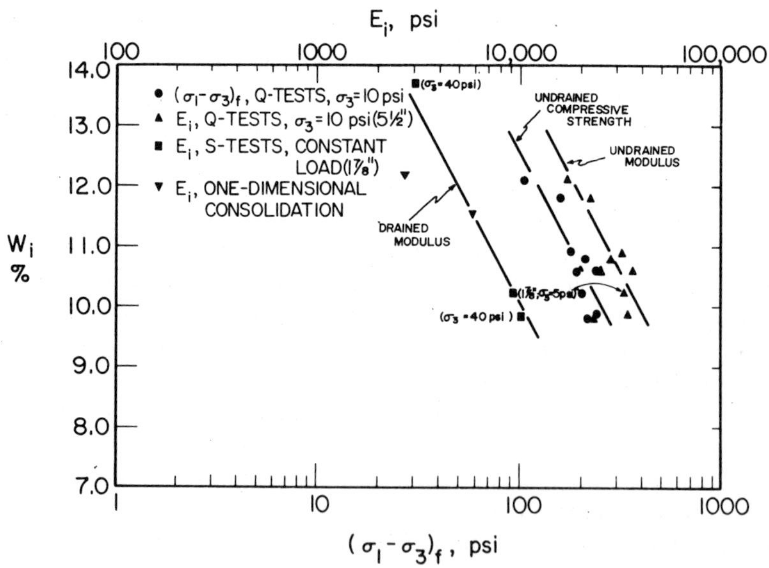

FIG. 12—*Modulus and compressive strength versus water content for black shale.*

determination of the water content or undrained strength. This relationship can be used in preliminary calculation of settlements at piers where the distribution of water content with depth is known but where drained test results are not complete.

A similar relationship between undrained strength, undrained modulus, and drained modulus for black shale is given in Fig. 12. In general, the black shale is stiffer than the gray shale. For the black shale the drained moduli are about one third of the undrained moduli.

In Figs. 13 and 14 the pressure meter test results are compared with the laboratory test values. The correlation was made by determining the water content from NX cores taken where the pressure meter tests were conducted. In Fig. 13 it is shown that the moduli from the pressure meter

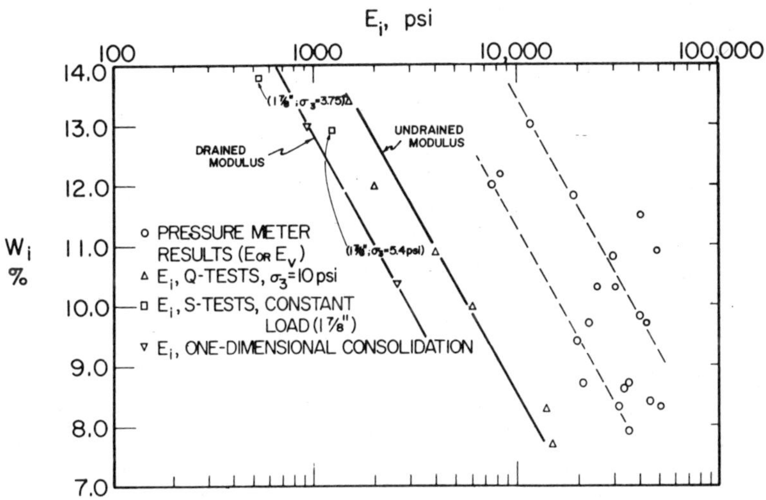

FIG. 13—*Comparison of laboratory and field modulus for gray shale.*

tests are about 3 to 7 times higher than the undrained moduli determined from the laboratory tests. About 90 percent of the pressure meter values fall within dotted lines roughly parallel to the relationship determined from the laboratory values. The values of undrained moduli from pressure meter tests on the black shale are shown in Fig. 14 and, generally, are about equal to the undrained values from laboratory tests, although there are a good number of points which show very high undrained modulus values from the pressure meter tests.

Conclusions

Laboratory results on the shales tested indicated that the undrained strength and undrained Young's modulus were related. The ratio of the

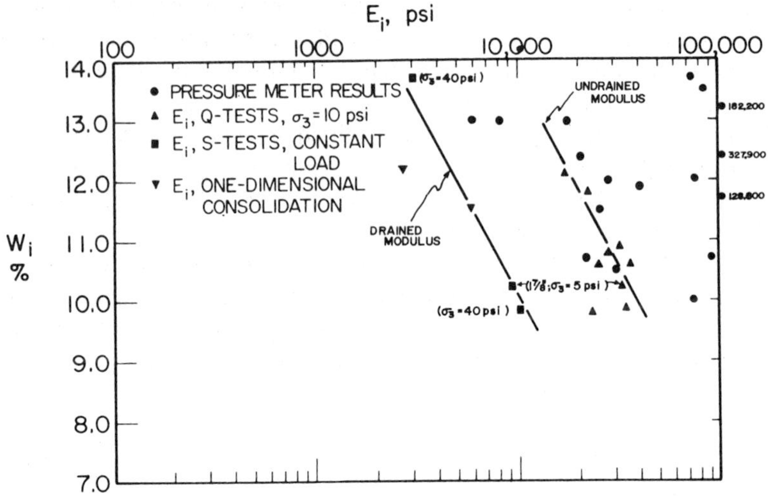

FIG. 14—*Comparison of laboratory and field modulus for black shale.*

undrained moduli to the undrained strength was 150:1 and 60:1 for the black shale and gray shale, respectively. It was also found that the drained Young's moduli for the same shales were about one third to one half of the undrained moduli for specimens with the same initial water content.

The undrained moduli, drained moduli, and the undrained strength of both shales were closely related to the initial water content. Thus the water content is an index property which can be used to estimate the modulus of shales at locations where complete test data are not available.

The deformation moduli from the pressure meter tests showed the same general variation with water content as did the laboratory tests. In all cases the moduli from the pressure meter tests were higher than those determined from laboratory tests conducted on specimens of comparable water contents. The pressure meter moduli are essentially undrained values, but, in general, are stiffer than laboratory undrained values, because of the possible disturbance of laboratory specimens, and because the loading in the pressure meter tests stresses the shale parallel to the bedding.

The moduli appropriate for design should be drained moduli, but the laboratory drained values are too low because of inevitable specimen disturbance. *In situ* pressure meter tests give values 5 to 13 times the laboratory drained moduli for the shales tested in this study. For purposes of design it is recommended that the laboratory drained moduli be tripled or that one third the pressure meter moduli values be used. In the absence of laboratory drained tests or *in situ* pressure meter tests, it is recommended that undrained moduli be used directly for calculation of settlements.

References

[1] Terzaghi, Karl and Peck, R. B., *Soil Mechanics in Engineering Practice,* 2nd Ed., J. Wiley & Sons, New York, 1967, pp. 84–88.

[2] Lambe, T. W., *Soil Testing for Engineers,* J. Wiley & Sons, New York, 1951, pp. 74–87.

[3] Menard, L., "Regles paur le calcul de la force partante et du tarrement des fondations en fonction des resultats pressiamelriques," *Proceedings,* Sixth International Conference on Soil Mechanics and Foundation Engineering, Canada, Vol. 2, 1965, pp. 295–299.

[4] Dixon, S. J. and Jones, W. V., *Civil Engineering,* CVEGA, Oct. 1968, Vol. 38, No. 10, pp. 34–36.

[5] Wilson, S. D. and Dietrich, R. J., "Effect of Consolidation Pressure on Elastic and Strength Properties of Clay," American Society of Civil Engineers Conference on Shear Strength of Cohesive Soils, Boulder, Col., pp. 419–435.

[6] Deere, D. U. and Miller, R. P., "Classification and Index Properties of Intact Rock," AFWL-TR-65-116, Air Force Special Weapons Center, Kirtland AFB, Albuquerque, N. M.

R. F. Coon[1] *and A. H. Merritt*[2]

Predicting *In Situ* Modulus of Deformation Using Rock Quality Indexes

REFERENCE: Coon, R. F. and Merritt, A. H., **Predicting *In Situ* Modulus of Deformation Using Rock Quality Indexes,"** *Determination of the* In Situ *Modulus of Deformation of Rock, ASTM STP 477,* American Society for Testing and Materials, 1970, pp. 154-173.

ABSTRACT: Field and laboratory test data from twelve construction sites show that the Velocity Index and the Rock Quality Designation (RQD) are indexes of *in situ* rock quality. The Velocity Index is the square of the ratio of the field seismic or sonic velocity to the laboratory sonic velocity. The RQD is a modification of core recovery measurements and can be correlated with the field static modulus of deformation. The Velocity Index has a similar, but poorer, correlation with the modulus of deformation because field dynamic tests involve a much larger volume of rock than field static tests. These correlations can be used to estimate the *in situ* modulus of deformation from core observations and geophysical testing.

KEY WORDS: tests, geological surveys, laboratories, rock, rock mechanics, modulus of elasticity, deformation, field tests, indexes (ratios), acoustic detection, evaluation

Large-scale field tests have shown that the modulus of deformation of a rock mass is lower than the static modulus of laboratory specimens taken from the same zone. It is now clear that the modulus is lower because laboratory specimens cannot duplicate the *in situ* conditions. While a laboratory specimen may adequately represent the rock material, it does not include the various rock defects such as joints and faults which have a profound effect on the deformation characteristics of the rock mass. Slippage and rotation of intact rock blocks along these geological discontinuities often create surface deflections that are larger than the deformation of the intact blocks themselves.

These observations have led to the obvious conclusion that engineering

[1] Geologist, A. A. Mathews, Inc., Arcadia, Calif.
[2] Geologist, H. G. Acres, Ltd., Niagara Falls, Ontario, Canada.

design must be based upon field moduli. Unfortunately, these tests are expensive and time-consuming, and only a small number can be performed on even the largest of construction projects. The selection of test sites and the proper application of test moduli to engineering design require a detailed geologic evaluation of the rock mass.

The geologic evaluation is normally a generalization of several diverse observations and tests. The character of the rock mass may be determined by field mapping, drill core, drilling records, *in situ* permeability tests, and geophysical testing. Variations in lithology, bedding, geologic structure, and the spacing and character of the geological discontinuities make this evaluation very difficult and essentially qualitative.

A recent investigation at the University of Illinois, Urbana, Ill., has examined the possibility of using these conventional geologic observations and tests quantitatively as indexes of the engineering properties of a rock mass. This paper presents two index properties; one based upon core logging, the other, on field seismic and sonic tests. It also presents relationships between these indexes and the modulus of deformation of the rock mass. The relationships are based upon data from twelve construction projects obtained through the cooperation of the U. S. Bureau of Reclamation, U. S. Army Corps of Engineers, the California Department of Water Resources, and several private companies. The study was conducted under the sponsorship of the Air Force Weapons Laboratory, Air Force Systems Command, Kirtland Air Force Base, N. M. 87117.

Field Investigation Methods

Predictions of the engineering properties of a rock mass may be based upon a wide variety of exploration techniques: core borings, permeability measurements, borehole photography, and geophysical surveys. This section describes the core logging and geophysical techniques used in this investigation.

Core Logging

Because core borings are one of the most common exploration methods, a system should be developed to obtain from these borings the maximum available information about the rock. Boring logs usually provides a brief lithologic description, a qualitative description of joints and faults, and the percentage of core recovered from each core-boring interval. Additional information often includes the type of drilling apparatus—single-tube, double-tube, or wireline core barrels—face or side discharge bits, and the size of stones in the diamond bit. These aspects of the drilling equipment may be important when boring logs are used to determine *in situ* rock conditions. In addition, the rate of drill advance can often indicate the abrasiveness, or hardness, of the rock.

Core Recovery—Core recovery is the most common method of expressing *in situ* rock quality. The core recovery is the percentage of the length of a cored interval represented in the core box by rock materials in any form, from cylinders to gravel; but, because both sound and weak rock are included in the core recovery, it is not a particularly sensitive measure of *in situ* rock quality. In basalt, for example, it is common to find brecciated zones between massive layers, and core recoveries of nearly 100 percent can be measured in both the massive and brecciated zones. In zones of heavily jointed rock, the core recovery may be high, whereas the core itself is broken into many pieces less than 1 ft in length. Low values of recovery are indicative of poor quality rock with fault zones or weathered areas. However, if the recoveries are high, say 80 to 100 percent, it does not necessarily mean that the rock can be considered uniform and massive.

Modified Core Recovery—The method of core logging developed by Deere [*1*][3] is a modification of the core recovery index called the Rock Quality Designation, or RQD. The RQD is the percent of the coring interval represented in the core box by sound, unweathered cylinders at least 4 in. long. Core fragments and weathered core are considered as additional core loss.

The RQD is a function of the length of individual pieces of the core; therefore, some discretion is necessary to determine whether the broken surfaces are the result of drilling or improper handling, or actually represent joint planes. The broken surfaces of the core can generally be described as:

1. Clean, irregular surfaces which can be rejoined with only a hairline separation.
2. Irregular and rounded surfaces that reflect considerable grinding in the core barrel.
3. Rather smooth surfaces that cannot be rejoined as in Category 1, but shows no signs of weathering (foliation planes, bedding planes).
4. Surfaces showing weathering, hydrothermal alteration, or slickensides.

Surfaces of Categories 3 and 4 are considered natural discontinuities of the rock. In this investigation the rock quality of each core run was expressed by the RQD.

Geophysical Testing

The geophysical measurements conducted in this investigation included continuous velocity logging in boreholes, up-hole and cross-hole seismic surveys, and seismic refraction tests on the ground surface and in tunnels.

[3] Italic numbers in brackets refer to the list of references at the end of this paper.

Data on dynamic moduli were also obtained from the locations of *in situ* deformation tests for purposes of correlating the static and dynamic properties of the rock mass.

The continuous sonic velocity measurements at the Dworshak, Two Forks, and Yellowtail sites were made by a three-dimensional sonic velocity logger owned by the Birdwell Co., Tulsa, Okla. Their logger was chosen for these measurements because it provides a continuous record of both compressional and shear wave arrivals. Birdwell also performed uphole, cross-hole, and refraction surveys using conventional seismic equipment. An Air Force Weapons Laboratory geophysical team, under the direction of Richard Zbur, provided additional seismic information at the Two Forks and Tehachapi sites. The remainder of the seismic data was obtained from the records of the Bureau of Reclamation, the Corps of Engineers, and the California Department of Water Resources.

Three-Dimensional Sonic Logger—The 3-D sonic logger is so named because the logs show the amplitude and time of arrival of the sonic energy for a given distance of travel. The 3-D sonde used in the NX boreholes is 2¼ in. in diameter, 12 ft long, and weighs 150 lb.

A sonic pulse is generated by a cylindrical, magnetostrictive-type transducer—a coil wrapped around a rod with magnetostrictive properties. The magnetic field caused by ac in the coil deforms the rod, producing a sonic pulse which travels into the rock. The pulse rate is approximately 20 repetitions per second and the resonance frequency of the transducer is 22 kHz. Cylindrical piezoelectric crystals are used as receivers and are located within the tool at distances of 3 and 6 ft from the transmitter. Acoustic isolaters between the transmitter and receivers prevent the sonic energy from traveling down the body of the tool. The arriving pulse deforms the crystals and produces the electricity necessary to record the energy arrival.

The arrival times are transmitted to the surface, amplified, and shown on an oscilloscope as an amplitude-modulated display. For a continuous recording of travel times that is more easily interpreted, the amplitude-modulated display is converted into an intensity, or variable density, display and presented as a series of dots and dashes. The intensity-modulated data are projected onto a 220-mm film whose movement is synchronized with the speed of the sonde in the hole, which is 30 to 100 ft/min.

The density of the rock and the variations in the hole diameter must be determined before the velocities can be calculated. For this reason caliper and gamma-gamma density logs are run in conjunction with the 3-D sonic logger.

Seismic Surveys—Seismic refraction surveys are used to determine the extent of surface weathering, the depth to a more dense geologic stratum, or the thickness of the zone of loosened rock (destressed zone) around an underground opening. Refraction surveys were used for this investigation to measure seismic velocities of the gneiss and schist in a tunnel at the Two Forks site and to determine the depth of surficial weathering at the tunnel portal. This method was also used to obtain velocities in the sandstone and the gneiss at the Tehachapi Pumping Plant site for correlation with the 3-D values.

The refraction survey measures the arrival time of seismic energy at a number of geophones placed at known distances from the shot point. The input energy depends upon the seismic velocity of the rock and the depth of penetration desired. For detailed work in a small area, an 8-lb sledgehammer striking a steel or aluminum plate is sufficient, whereas a blasting cap or dynamite is usually necessary for testing intervals of 100 ft or more.

The hammer-type seismographs used by the Bureau of Reclamation at the Glen Canyon, Morrow Point, and Two Forks sites have only one recording channel; thus the distance between the shot point and the geophone must be increased with each shot to get a complete record. A trip-switch on the hammer initiates the timing mechanism, and the first energy arrival at the geophone stops the counter. Most seismographs have a number of channels with one geophone each, which enables the entire survey to be run with one shot. The time of blasting must be synchonized with each of the recording channels so that the arrival times at the geophones can be measured. A blasting console is wired into the recording instrument, and the final record has the instant of the shot marked on the trace of each channel. As the energy arrives at each of the successive geophones, the trace shows a sudden increase in amplitude. A plot of time versus distance can be used to calculate seismic velocities and the depths to strata of greater density.

In an up-hole seismic survey a light charge is ignited at a given depth in a borehole, and the first energy arrivals are measured at a number of geophones in the same hole or on the surface. The charges can be located at various depths, but the initial shot is usually placed at the bottom, with successive ones at higher levels. Successive shots should not be made from the same position in the hole because blast-induced damage of the rock may result in anomalously low velocities. If the geophones are at the surface, it is recommended that they be in direct contact with the rock and at least 5 ft from the collar of the hole. If the geophones are in the same hole, the strength of the charge should be kept low to prevent damage to the geophones.

Velocity measurements can also be made between borings by using one hole for the shot and one or more holes for the geophones. For parallel shooting, the depths of the dynamite and geophones in adjacent holes are the same, while in fan shooting, the arrival time is measured at geophones at various depths in holes adjacent to the shot point.

Detailed measurements of seismic velocities in shallow borings or along tunnel floors require an instrument with a high-speed recorder. For example, to measure a 20-ft hole behind a jacking test in a tunnel, the first velocity readings should be obtained within 3 to 5 ft of the surface. If the rock is anticipated to have a velocity of 15,000 ft/s, the recorder must be able to measure travel times of 0.2 to 0.3 ms. At the present time, there are a few commercially made instruments with this resolution, since most geophysical apparatus is designed for oil field use. Consequently, the test requirements should be made clear to a geophysical company when contracting geophysical testing.

Laboratory Investigations

Laboratory tests were used to determine the engineering properties of intact rock, such as strength, modulus of elasticity, density, and compressional velocity. The tests were selected to provide laboratory data for correlation with the corresponding *in situ* properties. Because laboratory specimens represent rock in its most homogeneous and isotropic state, these comparisons provide a means of measuring the effects of geological discontinuities.

The test specimens used in this investigation were cylinders with a length to diameter ratio of approximately 2.0. They were cut from NX-size core (2¼-in. diameter) by a diamond saw, and the ends were smoothed to a tolerance of 0.002 in. by lapping.

Sonic Pulse Velocities

The compressional wave velocities of specimens collected at field sites were measured in both a saturated condition under successive uniaxial stress levels of 0, 50, 100, 200, 400, 1000, 3000, 1000, and 0 psi. The 3000-psi limit was chosen to avoid a specimen failure that could damage the velocity-measuring equipment.

Saturation is defined as the water absorbed by the specimen under the following laboratory procedures:

1. Oven-dry the specimen for 24 h at 105 C.
2. Place the specimen in a bell jar for 3 h under a partial vacuum of approximately 28 in. of mercury.
3. Cover the specimen with water while it remains under partial vacuum for another 3 h.
4. Allow the specimen to sit for 18 h under water at atmospheric conditions. This saturation procedure is based on the results of saturation tests presented in Table 1. These tests were performed on dense rocks, which should have long saturation periods, and show that a 3-3-18 h cycle provides the same percent absorption as the 24-24-48 used by Deere and Miller [*2*]. Following saturation the specimens were weighed and sonically

TABLE 1—*Percent absorption for several saturation cycles.*

Rock Type	Absorption, % Cycle Time, h 0-0-2	2-2-2	3-3-18	24-24-48
Granite Gneiss	0.09	0.11	0.14	0.14
	0.50	0.57	0.57	0.57
	0.11	0.25	0.25	0.25
Schist and Gneiss	0.12	0.20	0.19	0.20
	0.05	0.15	0.17	0.17
	0.08	0.08	0.10	0.10
Limestone	0.94	4.15	4.17	4.15
	0.19	0.62	0.62	0.62

NOTE—% Absorption $= \frac{P_2 - P_1}{P_1} \times 100$

where P_2 is the weight of saturated specimen and P_1 is the weight of oven-dried specimen.

tested. After being oven-dried for 24 h, their weight and velocity were measured again. In both sonic tests the ends of the specimens were covered with a thin film of silicone grease to improve acoustic coupling. The complete details of the pulse velocity apparatus are given by Deere and Miller, and the same equipment was used without modification for this laboratory program.

Unconfined Strength and Modulus of Elasticity

Unconfined compressive strengths and moduli of elasticity were determined for 280 specimens representing the entire range of rock types studied. Many of these specimens were taken from the sites of *in situ* jacking tests to compare the *in situ* and laboratory static moduli. The loads were applied by a 300-kip Riehle hydraulic testing machine and measured by a 600-kip capacity load cell. Axial strains were measured by two SR-4 electrical resistance strain gages mounted vertically on opposite sides of the specimen and wired in series. The load cell and SR-4 gages were connected to a Moseley X-Y Recorder, Model 2D, which plotted a stress-strain curve as each test progressed. The specimens were tested to failure at a loading rate of approximately 60 psi/s.

Evaluation of Test Results

Young's modulus, or the modulus of deformation, for rock is a function of the applied stress and may depart from the simple proportionality of stress and strain that applies to many engineering materials. The stress-strain relationship of intact rock specimens is often linear at intermediate stress levels and curvilinear at low and high stress levels. Deere and Miller suggest that the slope of the stress-strain curve at 50 percent of the unconfined strength, E_{t50}, be considered the characteristic modulus of intact rock. They feel that the curvilinear relationship at low stress levels is caused by the closure of stress-relief microfractures created during and after coring. The $E_{\overline{t50}}$ was used in this investigation.

The results of sonic velocity tests showed that in all cases the compressional velocities increased as the stress level increased and were higher for the saturated specimens than for dry ones. This has been observed by many researchers including Deere and Miller [*2*], Walsh [*3*], Birch [*4*], and Wyllie et al [*5*].

The velocities of the saturated specimen were chosen for correlation purposes because the 3-D velocities and most of the data from seismic surveys were obtained from saturated material. The selection of the proper stress level at which to measure the compressional velocities is important, because all rocks show an increase in velocity with pressure. Intuition might suggest that the stress level is governed by the overburden pressure acting at the depth from which the specimen was obtained. These velocities would be expected to be compatible with the 3-D sonic velocities at the same depth; however, the assumption that the pressure at a certain depth is equal to the overburden is not always justified. The *in situ* stress state is related to tectonic forces as well as the overburden pressure. In addition, it appears that a short-term test load equivalent to the overburden will not restore the specimen to its *in situ* condition. Therefore, it was considered most reasonable to select a stress level that resulted in the maximum increase in velocity for all rock types. All laboratory velocities used in this report were measured at 3000-psi axial stress. In the majority of cases the increase in velocity for the saturated rock had leveled off at this point. Triaxial tests of saturated rock cores up to 10,000 psi have shown a continual velocity increase throughout the entire test [*5*], although in many cases at least 90 percent of the velocity increase has taken place by 3000 psi.

Index Properties and Engineering Classification for *In Situ* Rock

The classification system introduced here is based upon two index properties of *in situ* rock, which are a measure of the discontinuities in the rock mass and were developed using (1) a modified core recovery technique (the Rock Quality Designation, RQD) and (2) the results of field and laboratory seismic and sonic velocity tests (Velocity Index). Two methods are therefore available for determining rock quality, and each provides a check on the accuracy of the other. The RQD, for example, is affected by the drilling procedure, whereas velocity measurements are not.

The Velocity Index is defined as the square of the ratio of *in situ* to laboratory, or intact, compressional wave velocities. The field velocities may be measured by in-hole sonic (3-D), refraction seismic, or cross- or up-hole seismic tests. The sonic and seismic data are normalized by the laboratory velocity in order to reduce the influence of lithology. If the field velocity is used to characterize rock quality, then a heavily jointed granite would have the same quality as an unjointed, friable sandstone. The ratio of the velocities in both low- and high-velocity rock will approach 1.0 as the number of joints decreases. The ratio is squared to make the Velocity Index equivalent to the ratio of the dynamic moduli.

Correlation or Rock Quality Indexes

Velocity measurements from up-hole seismic and 3-D sonic tests are correlated with the RQD in Fig. 1. To make these comparisons it was necessary to adjust the scale of the test intervals: the sonic logs, for instance, contain foot-by-foot velocity values, while the RQD is based on core runs of 5 to 10 ft. To compare these data, the sonic velocities were averaged for each core run. A Velocity Index, calculated from the average field velocity and a laboratory sonic test, is compared with the corresponding RQD value. The seismic tests, on the other hand, have test intervals of 10 to 75 ft, and it was necessary to compare the average RQD for the interval with a Velocity Index based on the field velocity and one or more laboratory tests.

Although the test interval is much larger for the up-hole tests, the points for both up-hole seismic and 3-D sonic tests generally have the same trend. An exception is that of the 3-D sonic tests at Tehachapi. These values are anomalous to the trend established by the up-hole seismic tests at Tehachapi and the seismic and sonic tests at other sites. For this reason they are not plotted individually, but their location is shown by a dashed line which would enclose 75 percent of the points.

The correlation coefficient r was determined by a reduced major axis analysis, because the RQD and Velocity Index are considered mutually independent measurements of rock quality. Least squares regression analysis requires the assumption of a dependent and independent variable and can therefore yield two different correlation lines and correlation coefficients. The reduced major axis analysis, on the other hand, yields one correlation line and one correlation coefficient. The significance level SL was determined from a table prepared by Fisher [*6*]. The significance level, which is determined from the number of data points and the correlation coefficient, indicates the probability that data points are part of a random distribution. A significance level less than 0.01 indicates a highly significant correlation.

Figure 1 is proposed as the basis of an engineering classification for *in situ* rock. The correlation of the RQD and Velocity Index presented there provides two alternate, and apparently compatible, methods for determining the character of *in situ* rock. Table 2 presents the classification with RQD and Velocity Index limits for Very Poor, Poor, Fair, Good, and

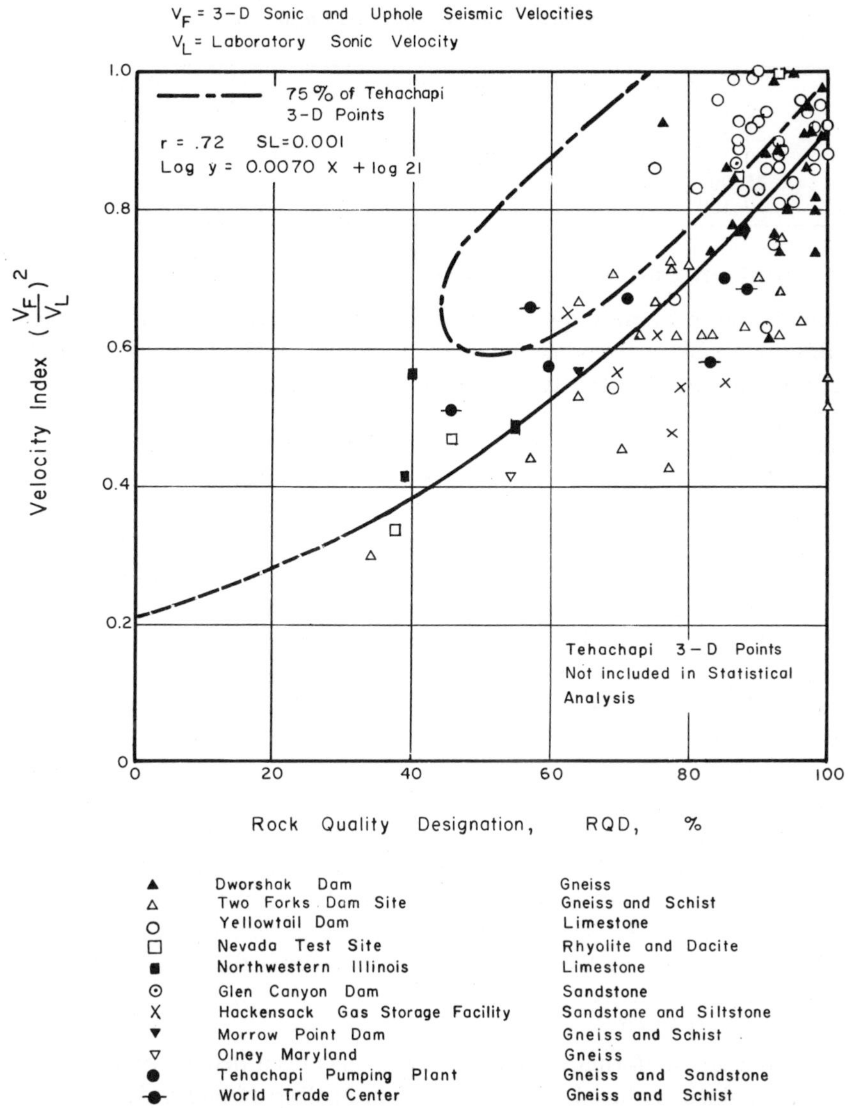

FIG. 1—*Relationship between RQD and Velocity Index for combined 3-D and up-hole data.*

Excellent rock. The RQD limits are those suggested by Deere [*1*], while the limits for the Velocity Index are based on the correlation line in Fig. 1.

At this point no quantitative engineering properties are implied by these descriptive categories. The usefulness of the classification can only be determined by comparing the measured engineering properties with the classification determined by the RQD and Velocity Index Methods.

TABLE 2—*Engineering classification for* in situ *rock.*

RQD, %	Velocity Index	Description
0–25	0–0.20	Very Poor
25–50	0.20–0.40	Poor
50–75	0.40–0.60	Fair
75–90	0.60–0.80	Good
90–100	0.80–1.00	Excellent

Deformation and Rock Quality Indexes

The influence of rock quality on the *in situ* modulus is well illustrated by the plate jack test moduli obtained at Dworshak Dam [7]. These tests offer a unique opportunity to determine the influence of rock quality, since the rock is lithologically uniform and the test site preparation and test procedure were essentially identical. A total of 24 tests yielded moduli ranging from 0.7 to 10 $\times$ 10^6 psi. This range includes most of the *in situ* static moduli reported in the literature.

It is apparent that increasing the number of tests at a site without some measure of the rock quality for each test increases the uncertainty in the modulus to be used in design. The use of an average modulus may be reasonable, but test values showing moduli about one fifth of the average might influence the engineer to apply a large factor of safety at some point in his evaluation. On the other hand, the use of the minimum modulus value for design could result in an overconservative design.

Rock quality measurements may be even more important than the comparison of *in situ* tests. If rock quality can be correlated with test moduli, then the rock quality measurements can be used to estimate the deformability of the rock mass at any location accessible by exploratory borings. This correlation would substantially reduce the number of *in situ* tests required while, at the same time, providing the design engineer with a much better picture of the modulus variation within the rock mass.

Deformation Modulus

Once the load-deformation curve (Fig. 2) has been defined by an *in situ* test, the engineer is faced with the problem of determining a design modulus from the test data. The test modulus is proportional to one slope on the load-deformation graph. It may be calculated from the slope of the curve from the origin to the point of total deformation and maximum load, E_{12}; the slope of the upper half of the unloading curve, E_{23}; or the slope of the entire unloading curve, E_{24}. While these moduli do not represent all possible interpretations, they represent those commonly used.

E_{12} is generally the lowest and, therefore, the most conservative value to use. It includes deformation during loading and under constant load. Depending upon the general shape of the load-deformation curve, this mod-

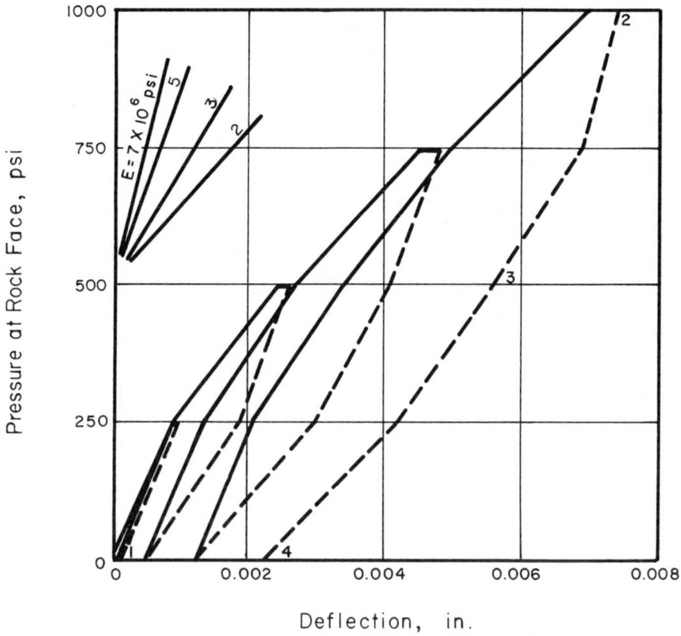

FIG. 2—*Load-deformation curve from plate jack test.*

ulus increases or decreases with increasing stress levels but is normally lower than E_{23} and E_{24}. This modulus will be referred to as the modulus of deformation, E_d.

The two moduli measured from the unloading curves can be taken at any cycle of the test; but as the loops are quite similar, they are commonly measured on one of the cycles to maximum load. E_{23} is the highest, and it has been noted that it often corresponds to the modulus calculated from seismic tests. E_{24} is lower, ranging in value between the seismic modulus and the modulus of deformation. Because E_{23} and E_{24} approach the *in situ* dynamic modulus they will not be further considered in this paper.

Comparison of RQD and In Situ *Moduli*

The RQD offers a method of determining rock quality which is independent of laboratory tests. As core borings are a part of the exploration program on most major engineering projects, a correlation of the *in situ* test results with the RQD could be made for just the small cost of core logging by the RQD method. The accuracy of the correlation between deformability and rock quality depends upon the accuracy of the rock quality determination at the test site. For this purpose, a 10 to 20-ft core boring drilled behind the center of the jack pad is recommended.

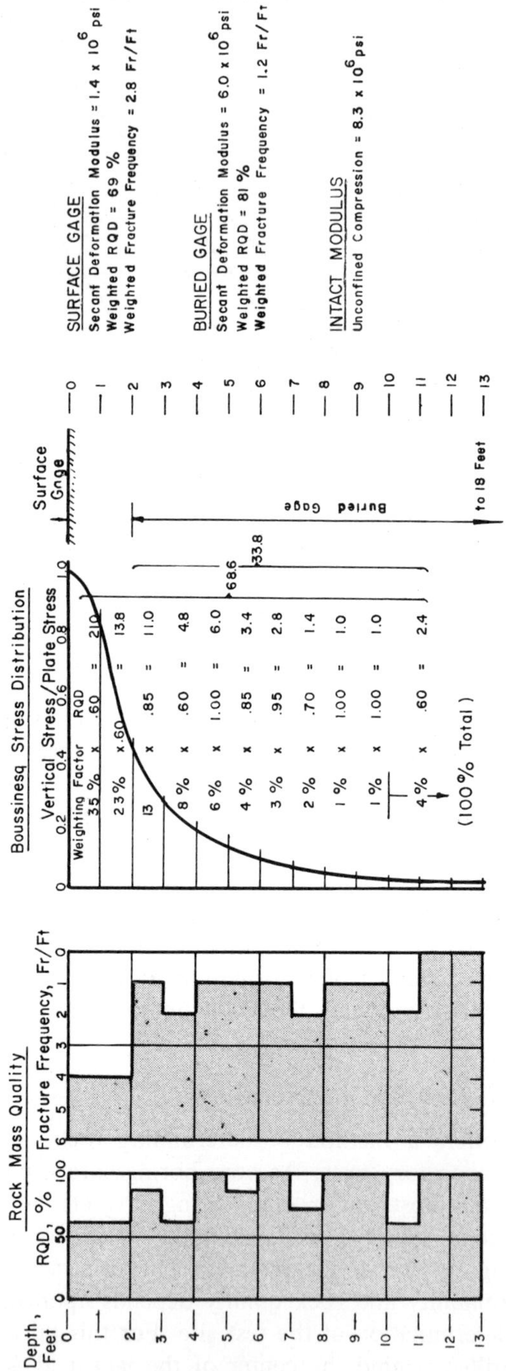

FIG. 3—*Illustration of a method for obtaining a weighted rock quality beneath a plate jacking test.*

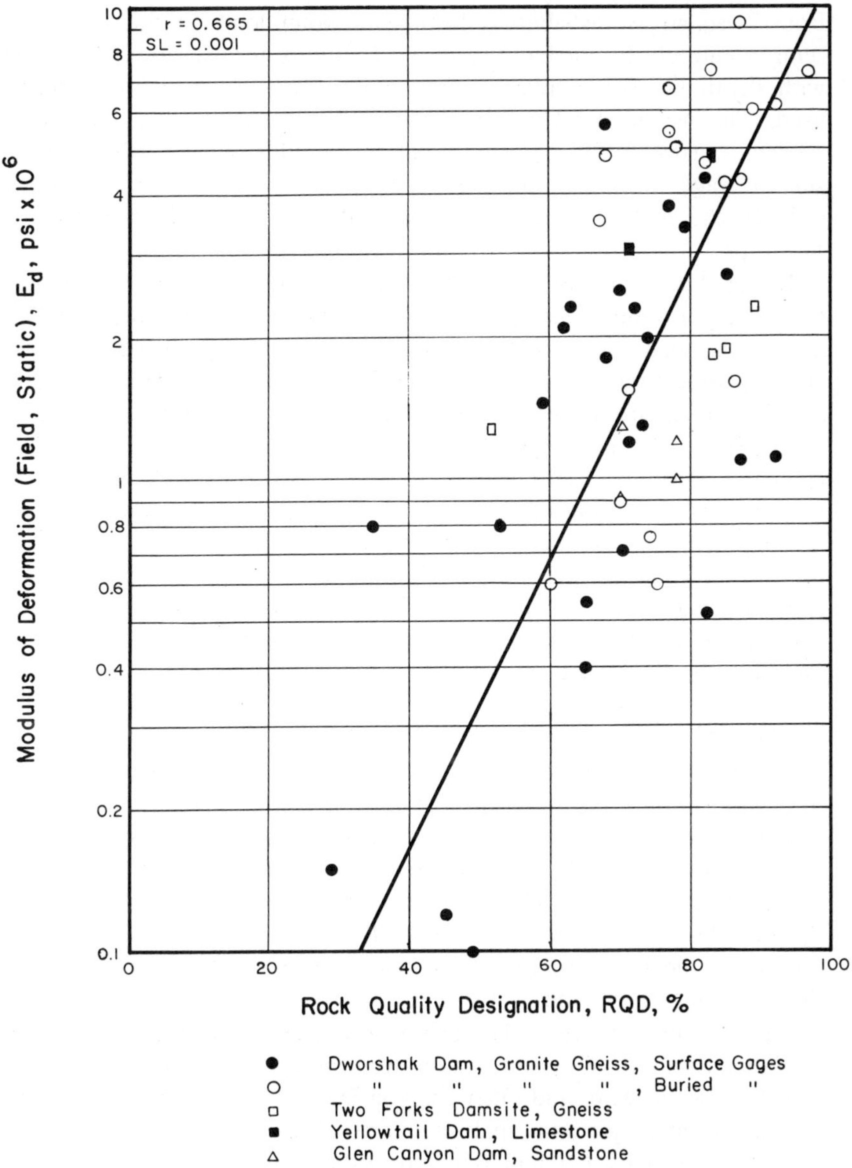

FIG. 4—*Comparison of the RQD and* in situ *static modulus of deformation.*

Theoretical studies show that changes in stress caused by a load at the rock surface decrease rapidly with depth; therefore, near-surface fractures have a more pronounced effect on the deflections measured than do deeper fractures, and for this reason the RQD measurements should be weighted according to depth. Our data have been weighted by the method of the

Boussinesq stress distribution beneath a point load on the surface of a semi-infinite elastic media. Figure 3 [*1*] shows the variation of the RQD with depth beneath a plate jack and illustrates the weighting technique used. The weighted RQD for the buried gage (depth 2 to 18 ft) is 81 percent, corresponding to a deformation modulus of 6×10^6 psi. For the surface gage determination the effect of fracturing around the opening reduces the RQD to 69 percent and the deformation modulus to 1.4×10^6 psi.

In Fig. 4 deformation moduli from the jack tests at Dworshak, Tehachapi, Glen Canyon, and Yellowtail are plotted against the weighted RQD. Although the relationship with a correlation coefficient of 0.665 looks promising as prediction chart, it should be noted that only gneiss, sandstone, and limestone are included. The intact moduli of the rocks range from 3 to 10×10^6 psi. Figure 4 could be used for deformation modulus estimates of similar rocks, but it cannot be used for rocks with low intact moduli. The limitation of this type of correlation is illustrated by the following example. Consider a sandstone with an intact modulus of 1×10^6 psi and an RQD of 80 percent. The relationship in Fig. 4 would indicate a deformation modulus of 2.8×10^6 psi, or nearly three times the intact static modulus. This estimate is obviously in error because the field static modulus should not be greater than the laboratory static modulus.

Comparison of the Rock Quality Indexes and Modulus Ratio

The limited application of a relationship such as Fig. 4 can be avoided by expressing the deformability of the *in situ* rock as a fraction of the deformability of an intact rock specimen. The ratio of the *in situ* modulus and the intact modulus is hereafter referred to as the Modulus Ratio. Although the intact modulus can be determined by either a static or a dynamic test, a Modulus Ratio defined by field and laboratory static tests appears to be the most logical choice. The intact static modulus used in the following discussion is the E_{t50}.

Figure 5 presents the relationship between the weighted RQD and the modulus Ratio E_a/E_{t50} for *in situ* tests at Dworshak Dam, Two Forks site, Yellowtail Dam, and Glen Canyon Dam. The best fit line, determined by a reduced major axis analysis, starts at an RQD of 100 percent and Modulus Ratio of 1.0 and descends rapidly to an RQD of 57 percent and Modulus Ratio of 0. The correlation coefficient is 0.544 for the value of the ordinate given by *y*. The best fit line shown is probably valid in rock with an RQD greater than 60 percent. Five data points with an RQD less than 60 percent indicate that the best fit line in the low Modulus Ratio zone would curve toward the origin of the graph.

Although Figs. 4 and 5 utilize the same test moduli and RQD, Fig. 5 has somewhat more scatter and, thus, a lower correlation coefficient. The

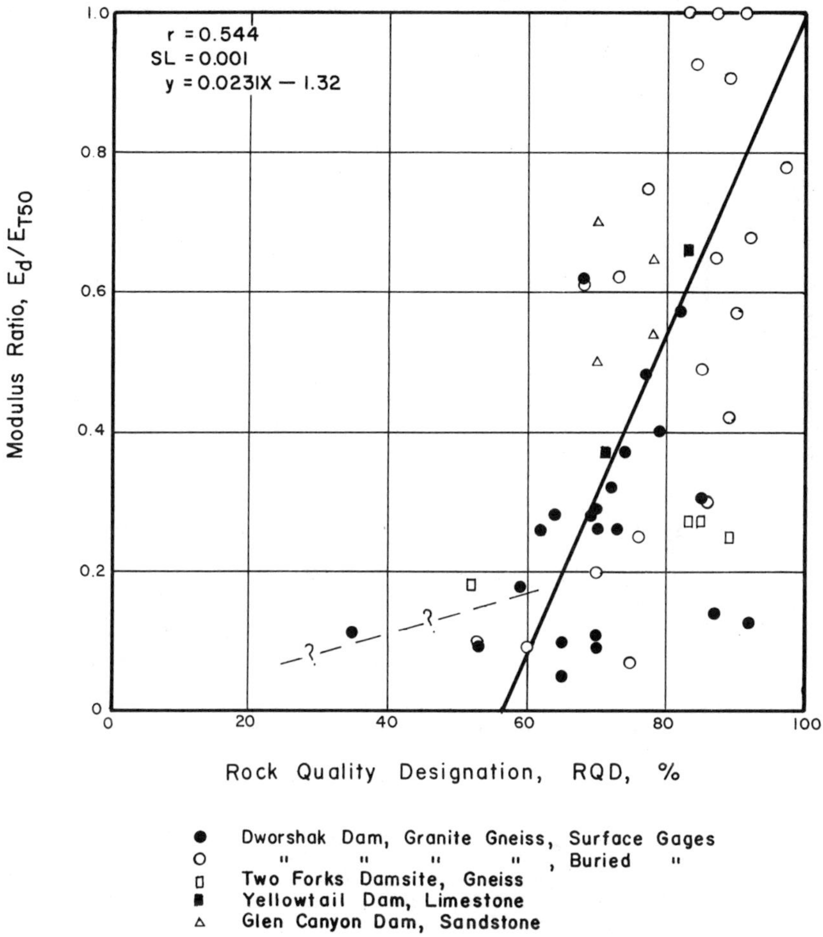

FIG. 5—*Comparison of RQD and the Modulus Ratio E_d/E_{t50}.*

scatter is very likely the result of using two or three laboratory tests (rather than one) to evaluate the intact modulus of each site; however, the increased scatter in this relationship is offset by a more universal application of the relationship. Note that the estimated modulus of deformation for a sandstone with an intact modulus of 1×10^6 psi and an RQD of 80 percent would be about 0.6×10^6, or approximately one-fifth the modulus estimate from Fig. 4.

Figure 6 graphs the relationship between the Velocity Index and Modulus Ratio for the same *in situ* tests. In this case the scatter of data points is much greater, and the correlation coefficient is 0.368. The best fit line now starts at a Velocity Index of 1.0 and Modulus Ratio of 0.8 and descends to a Velocity Index of 0.4 and Modulus Ratio of 0.

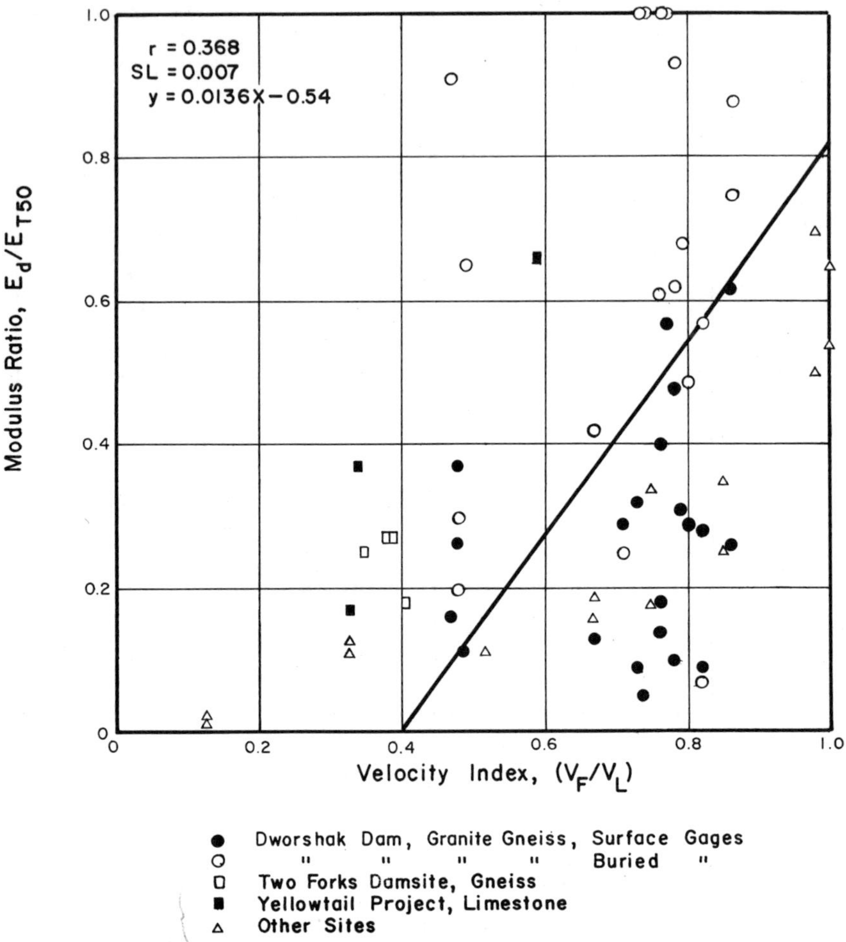

FIG. 6—*Comparison of Velocity Index and the Modulus Ratio* E_d/E_{t50}.

The poorer correlation could either be inherent in the relationship or indicate that the velocity measurements did not adequately measure the dynamic characteristics of the *in situ* test site. The latter explanation is probably more correct, as the volume of rock involved in the seismic tests was in all cases considerably larger than the rock stressed by the field static test. The relationship between the RQD and the Velocity Index (Fig. 1) suggests that the two rock quality indexes should have similar relationships with the modulus of deformation.

Three jack test borings at Dworshak were logged by the 3-D sonic method in an attempt to obtain dynamic measurements in a scale representative of rock loaded by the jack test. The results of this logging are not included in the Velocity Index versus Modulus Ratio graph because the first usable measurements were obtained when the sonde was 5 ft from the surface. As the first five feet are the most critical, this sonic logger is not adequate for this type of testing. A sonic sonde with a shorter transmitter-receiver spacing (possibly a foot long) might provide the kind of detailed dynamic measurements required. Seismic up-hole and cross-hole measurements using short test intervals and high-speed recording units might also provide the detailed dynamic measurements required.

Conclusions

The engineering classification of *in situ* rock presented is based upon two indexes of rock quality, the RQD and the Velocity Index. The RQD is a modified core recovery which includes only those pieces of sound core 4 in. or longer. The Velocity Index is the square of the ratio of the field velocity to the laboratory velocity. The field velocity is measured by either seismic or sonic tests; the laboratory velocity is measured by a sonic test on a saturated core cylinder under a 3000-psi axial load. According to the value of the RQD or Velocity Index, the *in situ* rock is described as Excellent, Good, Fair, Poor, or Very Poor. Correlations of *in situ* test data and the RQD and of *in situ* test data and the Velocity Index suggest that the engineering classification can be used to estimate *in situ* deformability. Possible applications of this method are (1) foundation modulus estimates for arch dams or heavy buildings and (2) estimates of the *in situ* modulus of rock outside the lining of a pressure tunnel. These results are summarized in Table 3 and in the following paragraphs.

TABLE 3—*Correlation of deformability and the engineering classification of* in situ *rock.*

Classification	RQD	Velocity Index	Deformability E_d/E_{t50}
Very Poor	0–25	0–0.20	> 0.20
Poor	25–50	0.20–0.40	> 0.20
Fair	50–75	0.40–0.60	0.20–0.50
Good	75–90	0.60–0.80	0.50–0.80
Excellent	90–100	0.80–1.00	0.80–1.00

A comparison of the *in situ* modulus of deformation and the RQD is presented in Fig. 4. Consistent trends in the graph suggest that the RQD can be used to predict the *in situ* modulus; however, the data used in these graphs represent a relatively limited range of *in situ* conditions. The graph, for instance, does not include tests on highly deformable rocks such as weak sandstones or shales with *in situ* moduli less than 1.0×10^6 psi. If such data were included, the RQD would probably not appear to have a unique correlation with the modulus. This possibility, which appears to be a very real limitation of the graph, can be eliminated by expressing the deformability as the ratio of the modulus of *in situ* rock to the intact, or laboratory, modulus.

This ratio, referred to as the Modulus Ratio, is calculated using the *in situ* modulus of deformation to define the field deformability and the laboratory static modulus, E_{t50}, to define the intact deformability. Comparisons of the Modulus Ratio with the RQD and Velocity Index are presented in Figs. 5 and 6. Each graph has about the same trend, but the Modulus Ratio versus RQD graph has a higher correlation. Both graphs show a rapid decrease in the Modulus Ratio to about 0.2 as the RQD decreases from 100 to 60 percent or as the Velocity Index decreases from 1.0 to 0.5. When the RQD is less than 60 percent or the Velocity Index is less than 0.5, the Modulus Ratio is generally less than 0.2, but the data in this range are too limited to define the relationship precisely.

The graph of Modulus Ratio versus Velocity Index has a greater scatter than the Modulus Ratio versus RQD graph, which is apparently due to a difference in the scale of the tests rather than a failure of the Velocity Index to measure rock quality correctly. For this reason it is suggested that the latter graphs be used to develop the relationship between rock quality and deformability and that velocity measurements be used in these relationships by first converting them to an equivalent RQD (Fig. 1).

Recommendations for Future Research

1. Nearly all the data presented in this paper are from rock classed from Fair to Excellent. Although zones of poor rock are commonly quite limited on construction sites, their presence may require significant design changes; thus it is important to extend the relationships to include Poor and Very Poor rock.

2. The Modulus Ratio versus RQD or Velocity Index relationships represent a limited sampling of rock types. Two of the sites are in gneiss or schist, one in sandstone, and one in limestone; additional studies of other rock types should be made to complete the relationships.

3. The correlations between the Modulus Ratio and the Velocity Index are inferior to those using RQD as the rock quality measurement. The correlation between the Velocity Index and RQD (Fig. 1) shows that the indexes are of about equal sensitivity. The scatter in the Modulus Ratio versus Velocity Index graph is probably the result of a difference in scale between the static and dynamic tests. The use of small-scale seismic or sonic testing at the location of the *in situ* testing should provide a Modulus Ratio versus Velocity Index relationship as sensitive as those developed from RQD measurements.

References

[*1*] Deere, D. U., Hendron, A. J., Patton, F. D. and Cording, E. J. in *Failure &*

Breakage of Rock, Proceedings of the Eighth Symposium on Rock Mechanics, American Institute of Mining and Metallurgical Engineers, Minneapolis, Minn., 1967, pp. 237–303.

[2] Deere, D. U. and Miller, R. P., "Classification and Index Properties for Intact Rock," Technical Report AFWL-TR-65-116, AF Special Weapons Center, Kirtland Air Force Base, N. M., 1966.

[3] Walsh, J. B., *Journal of Geophysical Research*, JGREA, Vol. 70, 1965, p. 399.

[4] Birch, F., *Journal of Geophysical Research,* JGREA, Vol. 65, 1960, pp. 1083–1102.

[5] Wyllie, M. R. J., Gregory, A. R. and Gardner, L. W., *Geophysics,* GPYSA, Vol. 23, 1956, pp. 459–493.

[6] Fisher, R., *Statistical Methods for Research Workers,* Hafner Publishing Co., New York, 1965.

[7] Shannon and Wilson, Inc., *Report on* In Situ *Rock Tests, Dworshak Damsite,* for U. S. Army Engineer District, Walla Walla, Corps of Engineers, Seattle, Wash., 1964.

R. E. Goodman[1]

The Deformability of Joints

REFERENCE: Goodman, R. E., "**The Deformability of Joints,**" *Determination of the* In Situ *Modulus of Deformation of Rock, ASTM STP 477,* American Society for Testing and Materials, 1970, pp. 174–196.

ABSTRACT: Recently developed procedures make it possible in cases with suitable geological conditions to assess the degree of safety and the probable displacements of excavations and foundations in discontinuous rock masses. In addition to geological data describing the geometrical system and mechanical data on the behavior of the rock, these analyses require data on the shear strength and deformability of the natural surfaces of weakness being considered. Several methods of sampling and testing joints and other discontinuities have been devised in Europe, and a large number of tests have been completed. Large shear machines have been used in Russia, England, Yugoslavia, and France for testing specimens up to 100 in.^{2} in area. In addition, a number of *in situ* block shear tests have been performed on naturally occurring discontinuities primarily by German, French, Yugoslavian, and Czechoslovakian geologists and engineers. The methods, results, and significance of these studies will be reviewed.

KEY WORDS: joints (junctions), geological faults, shear tests, rock mechanics, deformation, stiffness, drilling, sampling, evaluation, mechanical properties, foundations, rocks

Often, the displacements of rock portions of works must be known in order to complete engineering designs. Rock displacements arise from rock block deformation and adjustments between rock blocks.[2] In foundations, cuttings, and galleries sited in hard rock, the greater contributor to displacement is the latter, that is, closure and slippage of joints. Because joint sampling and testing is a recent development, the art is not fully appreciated and joint deformability values are not widely known. This paper presents a review of the methods of sampling and testing discontinu-

[1] Associate professor of geological engineering, Department of Civil Engineering, University of California, Berkeley, Cal.

[2] Discontinuities in rock include joints, sheared zones, crushed zones, seams, faults, cleavage planes, and openings along bedding, schistosity, and foliation. Joint is used here in the largest sense to refer to all natural discontinuities in rock.

ities and presents generalized deformability values for different types of discontinuities.

Methods for Producing Joint Specimens

In order to determine the mechanical properties of natural joint surfaces, it is necessary to test specimens of significant size in the laboratory. Various methods exist for producing these specimens, all of which will be discussed below. In some cases the natural joint surfaces are so regular, smooth, and planar that the creation of artificial joints by sawing of otherwise intact cores provides realistic specimens for analysis. Breaking beams in the field or laboratory and testing the artificial shear or extension features developed is another method possible here. One may also sample joints by selecting and mating blocks in an outcrop, carefully sculpting and removing them and then fitting them into a laboratory shear machine, or removing blocks containing joints by drilling, trepanning, or even wire sawing in the field. Finally, specimens may be obtained by drilling across joints in the field at angles normal, parallel, or inclined to the joints.

Artificial Joints

A large number of friction tests have been conducted in laboratories by sawing intact cores of rock in half, then mounting the two pieces in a direct shear box by setting in portland cement or epoxy or fitting them inside a triaxial device. In the latter case, the diamond-sawed cut must be at 45 to 60 deg to the ends of the core; but, even so, it is doubtful that such an unmodified cut is a very good model of a natural joint surface. Joints owe their origin to breakage of rock under extension or shear. Extensional joints are characteristically rough, and, while shear features are smooth, they often contain a filling of gouge, clay, or crushed rock. A diamond-sawed specimen can model a natural joint only if a gouge-like layer is developed on the diamond-cut surface. Research in this area is being pursued at the University of Illinois under the direction of Donald Deere of the Department of Civil Engineering. Diamond-sawed specimens are polished to a specified smoothness and then sheared at high normal pressure. This produces slickensides and crushed material on the joint surface, which resembles those sometimes seen on naturally occurring shear joints. The repeated cycles of loading of such specimens are then reasonable models of the actual *in situ* mechanical characteristics of the joint surface.

M. DeFreitas and John Knill of the Department of Geology at Imperial College, London, have been studying the shear behavior of artificially induced extension joints in granite. Rock blocks containing joints produced by the impact of simple beams, are brought into the laboratory from the quarry and the matching rock pieces are cemented into a mold fitting the shear box. Joints produced and tested in this manner are typically rough extension joints with a waviness of several centimeters that are very realistic models of actually occurring joints of extension origin. Artificial joints may also be produced in the testing machine by incipient shear failure of intact specimens. Both triaxial and direct shear specimens can be used.

Field Sampling of Open Joints

Joint surfaces in the field can be so open that one block rests on another with perfect fit but without any cohesion, in which case, it is possible to remove blocks containing joints with a geology pick or a crowbar. Thus, it may not be necessary to hold the upper and lower blocks together for shipment, but only to mark their relative positions so that they may be properly oriented in the laboratory. D. Krsmanovic and his associates in Sarajevo, Yugoslavia, have sampled a large number of joints in limestone by this method [*1, 2, 3*].[3] Of course, the samples are disturbed in the sense that any original closing of aperture tangentially or normally that may have occurred during the actual loading in the field cannot be reproduced in the laboratory specimen.

Drilling Joint Samples

It is possible to obtain a relatively undisturbed sample of a joint by drilling through it with a core barrel which may be normal, parallel, or inclined to the plane of the joint. The firm of Coyne and Bellier in Paris, France, has obtained samples at several damsites by drilling perpendicular to joint planes so that the joint occupies a position parallel to the ends of the core; then, after trimming in the laboratory and potting in portland cement, the circular cross section joint surface is tested in a direct shear machine. Samples from 2 to as much as 12 in. in diameter have been obtained in this manner (Fig. 1). But it is possible that drilling across the joint by a rotary drill will seriously disturb the joint surface, since one block may rotate past the other; in fact, it might be impossible to sample very weak or open joints by this technique.

A modification of the perpendicular drilling technique has been used at the University of California, Berkeley. In this method, a small hole is first drilled across the joint; then, a rock anchor is installed and tightened; and, finally, the small hole is over-cored—the sample is obtained without relative movement between the two blocks.

Coyne and Bellier, the Portugese National Civil Engineering Laboratory, and the mining group at Imperial College have drilled samples parallel to the plane of joint features so that the joint specimen occupies a diametral

[3] Italic numbers in brackets refer to the list of references at the end of this paper.

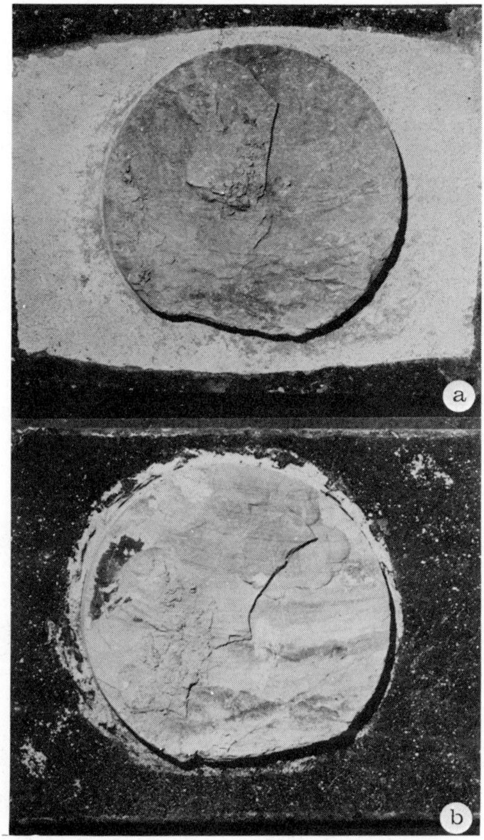

FIG. 1—*Large direct shear specimens of a natural joint obtained by drilling 12-in. (300-mm) cores across the joint.* (a) *and* (b) *are the bottom and top surfaces, respectively, after the test. Courtesy Coyne and Bellier.*

plane of a core sample. Figure 2 shows such specimens after a test in which NX size ($2\frac{1}{8}$-in.) specimens were encased in portland cement. Coyne and Bellier have also drilled larger specimens (Fig. 3*a*). David Pentz at Imperial College, England, has used a 9-hp diamond drill (Fig. 4) to obtain cores 9 in. in diameter. The cores were drilled without a core spring, and, when the full depth of drilling had been reached, the diamond bit was removed and a core spring and dummy bit were installed. The sample was then pulled by the core spring, jacking the drill against the collar, to produce a tension break at the end. The core containing the joint specimen could then be removed from the hole.

After removal, the joint is bound so that it will not be disturbed in transit and parafinned (Fig. 3*b*), or otherwise protected, so that the natural moisture content will be preserved. On return to the laboratory, the core

FIG. 2—*Joint specimens obtained by drilling NX cores parallel to the joint plane. Photos taken after testing of* (a) *lower and* (b) *upper halves. Note clayey filling material. Courtesy Coyne and Bellier.*

FIG. 3—*Large (10-in. diameter) core specimens containing joints in the diametral plane*: (a) *after removal from the hole;* (b) *packed in paraffin to preserve moisture content. Courtesy Coyne and Bellier.*

is fitted into the shear box by imbedding it either in concrete (Coyne and Bellier method) or in epoxy (Imperial College method). A surprisingly large number of samples have been obtained in this way, with dimensions up to 1600 cm^2.

It is also possible to drill specimens for triaxial joint tests by orienting the drill at 45 to 60 deg to a naturally occurring joint. A triaxial test with a naturally occurring joint at this controlled orientation is called a multistage triaxial. Tests of this type were performed on granitic rock by the Missouri River Division of the Corps of Engineers [*4*], on marble by Heuze and Goodman [*5*], and on a number of rock types by Jaeger [*6*] in Australia.

FIG. 4—*Large direct shear specimens obtained by drilling 9-in. cores parallel to selected joint surfaces. Courtesy Imperial College, Royal School of Mines.* (a) *specimens as received from field;* (b) *after removal of protective jacket.*

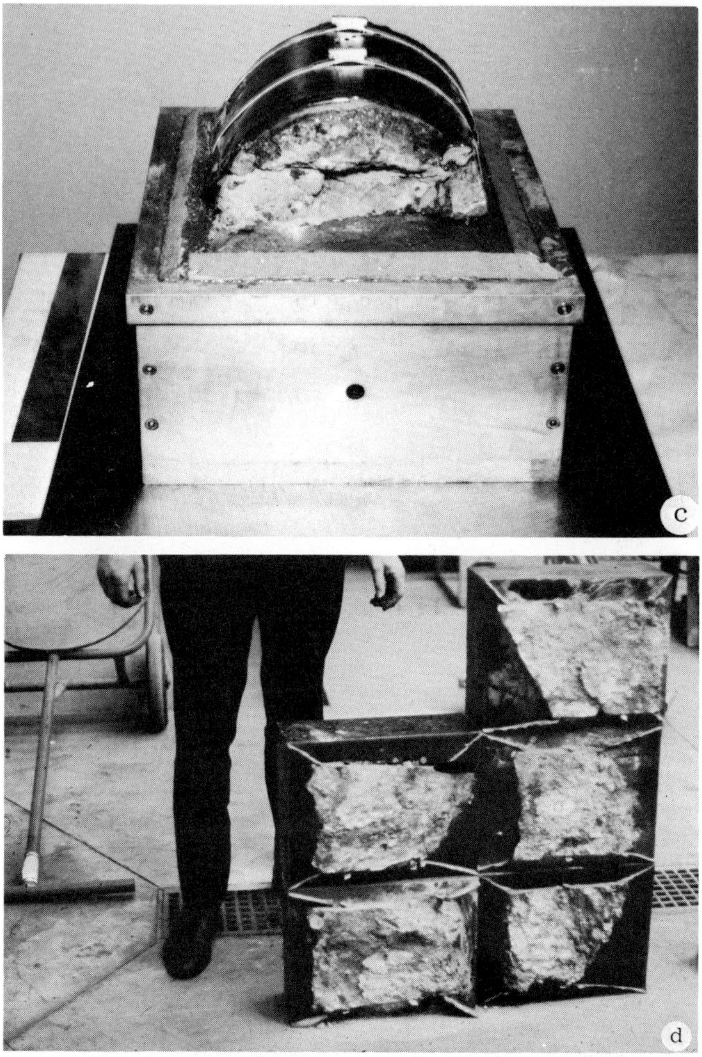

FIG. 4—*Continued.* (c) *mounted in shear box, ready for testing* (*confining bands will be cut before test*); (d) *specimens after test.*

Removal of Blocks Containing Joints

It is possible to remove large chunk samples of undisturbed rock and thus obtain large numbers of specimens for the laboratory. Such samples can be obtained by drilling overlapping holes with a drilling template to create a series of slots: for example, the U. S. Bureau of Reclamation took several large chunk samples from the Grand Coulee Third Powerhouse by

FIG. 4—*Continued.* (e) *cross section through a specimen showing rock clips used as filler in epoxy.*

this technique. The Bureau of Reclamation, the U. S. Army Corps of Engineers, and the Portugese National Civil Engineering Laboratories all have field diamond saws which can be used for removing large chunk samples. The firm of Coyne and Bellier has introduced a wire sawing apparatus consisting of two rods with pulleys at each end which are fitted into the drill holes. The wire runs between the pulleys, fixed into two corner holes of the eventual block and fed out of the drill holes to the continuation of the wire loop. In principle, four slots could be cut in this fashion to outline the four walls of the block; but, in practice, several of the sides are trepanned. Then the bottom of the block is freed by sawing with an ingeniously contrived loop in the wire. With the wire-cutting procedure it is possible to remove relatively undisturbed samples of soft weaknesses in hard rock. Eurenius and Fagerström [7] of the Swedish firm Vattenbyggnadsbyrån have taken a large number of block samples at a Syrian damsite in soft rock by cutting with a power saw. Direct shear specimens were prepared in the laboratory by normal drilling of the seams contained within the block samples. Other methods of cutting blocks which may eventually prove to have application for taking block samples are electron beam cutting, laser cutting, and, possibly, hydrogen arc cutting.

Methods of Testing Joints

Triaxial Test with Oriented Joint—Multistage Triaxial

As stated above, it is possible to conduct a triaxial test with an oriented joint so that the eventual failure is through the mode of slipping along a preselected surface. The specimen is failed under some initial confining pressure that is followed by a step-wise increase of confinement and sliding by increasing axial pressure. This yields a value for the joint angle of friction. The test may be reasonable in stiff joints where there is no filling or compressible rock zone between the walls of the joint. I [*8*] performed a finite element analysis of triaxial specimens containing inclined joints more compressible than the surrounding rock. The results indicated very uneven stress distributions along the joint and totally different concentrations of stress in the upper and lower walls of the joint; thus, the multistage triaxial is probably not a good test for seams, or joints of shear origin.

Laboratory Direct Shear Tests

The direct shear test is a useful method for evaluating the relative sliding tendency of joint blocks in an excavation or foundation; however, it is not a simple matter to produce a reasonably uniform stress distribution inside a direct shear box. If joints were perfectly smooth, it would be possible to align the joint inside a direct shear chamber and displace the top half relative to the bottom by means of colinear forces. Ideally, these forces would be shears along the top and bottom blocks, the blocks being very thin. However, since the top and bottom blocks must have some thickness for the specimen of rock to be sufficiently stiff, it is not practical to force the top over the bottom by shears along their surfaces. It is customary, instead, to push or pull the top relative to the bottom by applied forces and reactions transmitted from the sides of the block. These forces introduce overturning moments into the shear box, which must produce reaction moments from its top and bottom to effect a nonuniform distribution of normal pressure along the sliding surface. One possible result is indicated in Fig. 5. If the joint surface is rough, it is necessary to provide a sufficient gap between the top and bottom supporting blocks to allow for the full amplitude of the irregularity of the joint surface. This introduces a further overturning moment and a tendency for a nonuniform stress distribution in the shear box. There are additional moments in the vertical plane, introduced by hinging around high points of rock along the sliding surfaces.

Another problem of direct shear tests concerns the lateral boundary conditions. In evaluating the strength of rock joints for purposes of slope stability analysis, it is reasonable to conclude that the prototype blocks

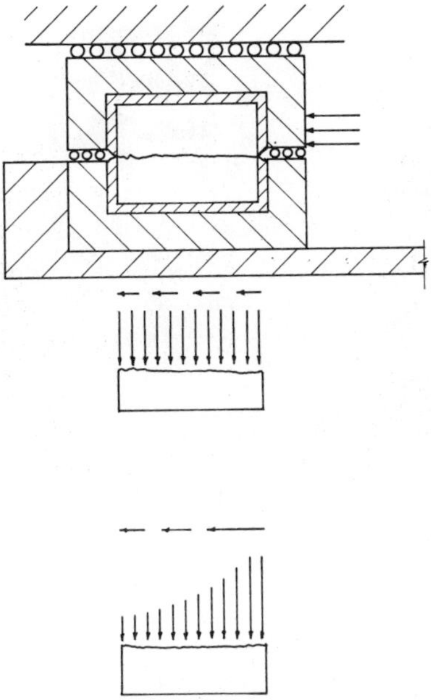

FIG. 5—*Stress distribution in a direct shear box.*

will be in a condition of plane strain, that is, translation is unaccompanied by rotations in the joint plane. This is the boundary condition usually built into direct shear machines distributed in soil mechanics laboratories around the world. With irregular joint surfaces, rotational tendencies are possible in the horizontal plane because of rotary moments on high points along the joint surface; in other words, the upper or lower block tends to swing into the sides. In the plane strain type of shear box, this is impossible because of the lateral confinement offered by the sides. However, the resulting distribution of side pressures is very irregular and cannot be known; the test results may therefore be unreal. Another approach to this problem is to omit the sides of the shear box altogether. The problem of achieving a uniform stress distribution in a direct shear box is difficult. A thorough study by modern analytical methods has not been made. Shear strength characteristics determined in direct shear tests depend on the boundary conditions achieved. Unfortunately, there is no uniformity in the construction of machines in use around the world; thus, the results of one worker may not agree with the results of another even though identical specimens are tested.

One of the most elaborate direct shear machines in the world was developed by Hoek and Pentz and their colleagues [*9*] at the Royal School of Mines at Imperial College, London, England (Fig. 6). In this machine, the horizontal force required to maintain the displacement in a single

1 — *Lateral load cells*
2 — *Trolley holding shear box*
3 — *Lateral restraining yokes*
4 — *Spherical seating*
5 — *Yoke for normal load*
6 — *100 ton screw jack for tangential load*
7 — *Tangential load cell*
8 — *Spherical seating*

FIG. 6—*Large direct shear machine at Imperial College, Royal School of Mines.*

direction can be controlled and measured. The shear box takes specimens up to 9 in. in width. Controls for pressure and rate of displacement have been incorporated in its design. Unfortunately, it is so recent a development that relatively few samples have been tested as of this date.

Plane strain shear boxes, designed for soil mechanics use, have been adapted for joint samples in many laboratories. In addition, similar shear boxes of much larger size and load capacity have been constructed for work on rock specimens by DeFreitas and Knill at Imperial College, by Deere and Coulson at the University of Illinois, and by the Societé d' Equipment Industrial et de Laboratoire SEIL in Paris. It is desirable in a direct shear machine that the normal load follow the position of the center of the contacting area at all times. The SEIL machine, which takes specimens with dimensions of 30 by 50 cm or less, is provided with hinges that allow the normal pressure ram to move with the relative displacement across the joint. Movement of the axis of normal load with the moving block is accomplished in the other machines mentioned by employing dead weights or hangars.

D. Krsmanovic [*1*] has constructed a large capacity machine without sides (Fig. 7). Specimens for this machine have dimensions of 40 by 40 by 20 cm, giving a shear surface of about 1600 cm^2 (the maximum shear stress is 80 kg/cm^2 on these large areas). The Krsmanovic device is designed such that the applied shear force is inclined about 4 deg to the surface of shearing. Inclination of the applied force is one way of overcoming some of the difficulties with the rotary moments described above; however, a large inclination of the applied force causes normal stresses to vary appreciably with the shear stress applied, which makes it impossible to fail specimens under low normal stress. The Krsmanovic device uses jacks on both sides of the ball bearings both above and below the shear blocks so that the center of the contacting surfaces never move during the test. Therefore, it is possible to use a fixed loading frame to apply the normal pressures.

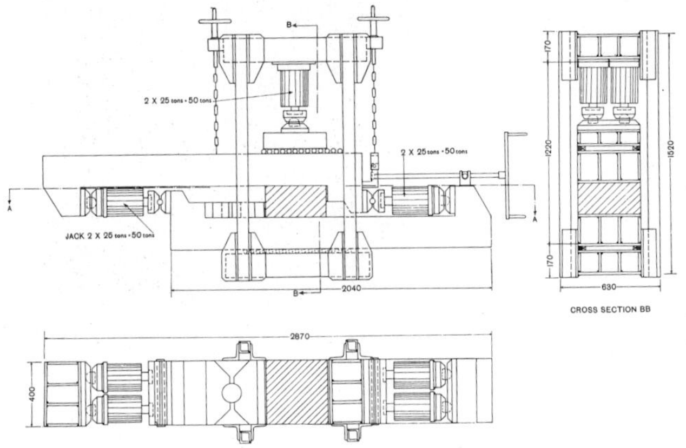

FIG. 7—*Schematic diagram of Krsmanovic shear machine.*

In Situ *Direct Shear Tests*

It is possible to perform direct shear tests in the field. These tests are carried out on a block formed by selective excavation of a natural surface of weakness that need not be horizontal. There are many ways of performing this test.[4] For all of them the block is usually confined in a reinforced concrete or a composite steel concrete frame. The shear load is slightly inclined, while the normal load is sometimes provided only by the block's own weight. Blocks on inclined surfaces are under shear stress even before the test begins. To extend the elastic information of the test, the block may be pushed up the hill, through zero.

Few tests of this type have been performed on hard rocks in the United States, owing to the expense in labor required. Many have been conducted by European organizations, but relatively few on natural joint surfaces. An example is shown in Fig. 8.

[4] Standards for *in situ* shear tests are being presently considered by ASTM.

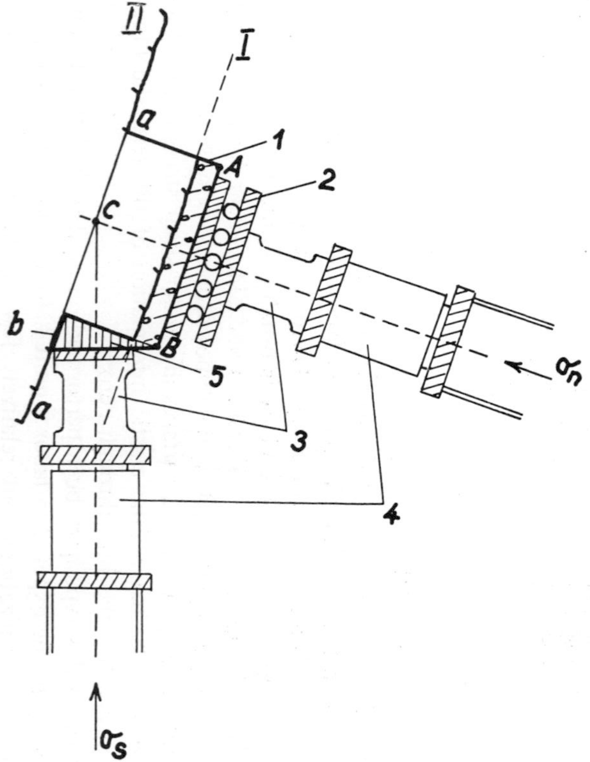

Shear test on a plane of mechanical discontinuity with shear movement within the rock mass.
1) *Concrete block with measured points A and B;*
2) *Roller bearing;*
3) *Strain gauge dynamometers;*
4) *Hydraulic presses for pressures of 50 or 100 Mp;*
5) *Plaster levelling wedge: a) Broken-off rock round the block; b) Cardboard lubricated with vaseline; c) points at which the movement on a plane of mechanical discontinuity is measured; I and II) planes of mechanical discontinuity.*

FIG. 8—In situ *shear tests conducted in Czechoslovakia.* (a) *conventional test on a horizontal plane;* (b) *test on a steeply inclined plane. Courtesy Karel Drozd.*

Joint Stiffness Properties

Joint deformability can be described by the character of the stress-deformation curves in direct shear. The terms "shear stiffness" and "normal stiffness" refer to the rate of change of shear stress with respect to shearing displacement and of normal stress with respect to normal displacement, respectively. Together with values for peak and residual shear displacements and maximum joint closure values, these quantities allow computation of the contribution of joints to the total displacements of structures in rock. Since the peak shear stress and displacement are known to be dependent upon the normal stress and joint roughness, filling, and other joint system variables, one may conclude that for a given type and condition of joint the shear stiffness is a function only of the normal stress. The normal stiffness, which depends exclusively on the joint system—initial aperture, roughness, type and character of filling, etc.—is zero under tensile loading; it increases under compressive loading to a very large number as the joint approaches a fully closed condition. Numerical values for these properties, revealed by tests on joint specimens, are reviewed below.

Results of Shear Tests

Results from a large number of laboratory direct shear tests and several *in situ* block shear tests have been summarized in Tables 1 and 2.[5] These data were gathered mostly from unpublished reports generously made available by the organizations concerned.

The load-deformation curves for these tests have been represented by values for shear stress and displacement at the peak and residual points. In some cases, a yield point was reported in the beginning part of the curves. Though the possible combinations of values for these quantities are numerous, several generalizations can be made about the characteristics of the results for different types of weakness surfaces. The load-deformation curves were classified into four types, drawn in Fig. 9 and summarized in Table 3.

1. Type 1 stress-deformation curves rise steeply to peak stress at very low deformations and then quickly fall to a residual value that may be one third or less of the peak. Healed joints and incipient fractures belong to this class.

2. Type 2 curves also develop peak stress at very low deformations but do not fall so sharply to the residual value; furthermore, the peak strength is only slightly above the residual strength. Artificial joints formed by diamond-sawed rock specimens belong to this class. Polished surfaces,

[5] TABLE 1—*Results of laboratory shear tests on rock specimens with joints.* and TABLE 2—*Results from* in situ *block shear tests on weakness planes.* are available from University Microfilms, Inc., 300 N. Zeeb Road, Ann Arbor, Mich. 48106, $3.00.

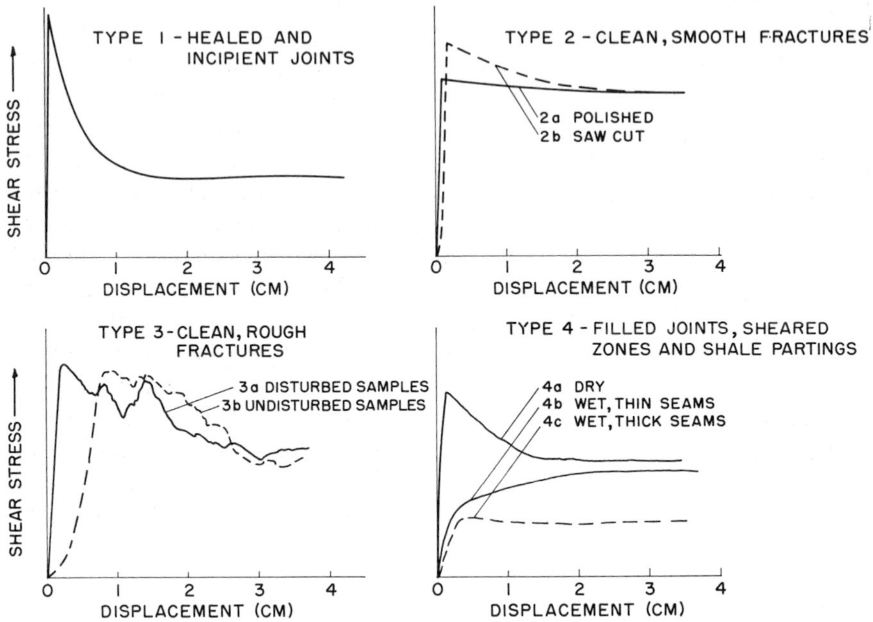

FIG. 9—*Typical shear stress-deformation relationships for various weakness surfaces.*

Type 2a, show less rate of decay of strength after the peak than do unpolished saw cuts, Type 2b. The shear stiffness is higher than in natural joints, but is still low enough in value (10^4 kg/cm^3) to contribute significant deformation beyond that effected by an intact hard rock whose joints are repeated at intervals of several feet, such as quartz monzonite.

3. Type 3 curves show lower stiffness and numerous secondary peaks, which are sometimes as large as the primary peaks. These curves result from the testing of clean, rough joints such as extension joints in granite. The irregularities in the load-deformation curve result from overriding of successive asperities. Undisturbed samples of such joints may show an initial concavity as the tangential aperture is closed (Type 3b).

4. In Types 1–3 it mattered little whether the surfaces were dry or soaked with water; in Type 4 curves, on the other hand, the behavior is modified grossly by a change in the water content. These curves are typical for sheared zones, clay-filled joints, and shale partings. At low moisture contents (Type 4a), the curves show high stiffness and great difference between peak and residual displacements, although generally not as severe as for the healed joints (Type 1). When wet, the stiffness is greatly reduced. Thin zones (Type 4b), when wet, show "strain hardening" behavior, the ultimate strength being considerably greater than that at the point of

TABLE 3—*Typical load deformation relations for different types of seams and fractures.*

Type	Type of Surfaces Typically Exhibiting Behavior in This Class	Peak Displacement, cm	Tangential Stiffness, kg/cm³	Peak Strength ÷ Residual Strength
1	Healed joints and incipient joints	0.03	6.0×10^2	3.0
2	Clean, smooth fractures			
	2a polished	< 0.01	6.0	1.1
	2b unpolished (rough saw cut)	0.02	1.3	1.2
3	Clean, rough fractures			
	3a artificial extension fractures and disturbed samples	0.20	1.3	1.75
	3b undisturbed samples	0.80	1.0	1.75
4	Filled joints, sheared zones, shale partings, and smooth bedding			
	4a dry or slightly moist	0.12	2.2	1.5
	4b wet; thin seam	0.26	1.5	0.6
	4c wet; thick seam	0.30	0.8	1.0

TABLE 4—*Results of laboratory direct shear tests on artificial specimens with mica-filled joints.*

Test Number	Amplitude of Roughness, cm	Thickness of Seam, cm^2	Cross Section Area, cm^2	Normal Stress at Start of Test, kg/cm^2	Tangential Stress, kg/cm^2		Tangential Displacement, cm		Shear Stiffness, kg/cm^3		Curve Type
					Yield	Peak	Yield	Peak	Yield	Peak	
1	0.41	0	31.7	7.6	...	6.18	...	0.203	...	30.4	4a
2	0.41	0	31.7	7.6	...	6.80	...	0.244	...	27.9	4a
3	0.41	0.1524	31.7	7.6	...	4.33	...	0.241	...	18.0	4a-4b
4	0.41	0.524	31.7	7.6	...	4.26	...	0.226	...	18.8	4a-4b
5	0.41	0.3048	31.7	7.6	3.05	3.28	0.102	0.165	29.9	19.8	4b-4a
6	0.41	0.3048	31.7	7.6	...	3.06	...	0.305	...	10.1	4b-4a
9	0.41	0.3048	31.7	7.6	1.75	2.58	0.114	0.297	15.4	8.7	4b-4a
7	0.41	0.508	31.7	7.6	1.43	2.05	0.140	0.291	10.2	7.1	4b
8	0.41	0.508	31.7	7.6	1.83	1.98	0.203	0.241	9.2	8.2	4b

maximum curvature (denoted as "peak"). Thick seams may have elastio-plastic stress-deformation curves when wet (Type 4c). All of the *in situ* tests (Table 2) yielded Type 4 curves, probably because these expensive tests are reserved for the worst discontinuities.

It is difficult to increase the water content of sheared zones and clay seams to desired values. The field moisture content can be retained by careful sealing in the field; but, to increase the moisture content beyond this value, soaking is necessary. Soaking the specimens without surcharge may lead to excessive moisture contents well beyond those expected in the field. Since the permeability is low under normal pressure, soaking under surcharge tends to wet the edges more than the center. Coyne and Bellier have tried introducing water into drill holes parallel and perpendicular to the seam with moderate success.

Typical values for the peak displacement, tangential stiffness, and ratio of peak to residual stress are given for all types of surfaces in Table 3. It is emphasized that these are only typical values; considerable deviation exists in all these quantities. Both the peak displacement and shear stiffness tend to increase with increasing normal stress. When variation in normal pressure is ignored, the deviation of peak displacement is, however, far less than the deviation of stiffness. In fact, if no test were possible, as a rough estimate of initial shear stiffness, one could calculate the strength from the friction angle (which is usually between 30 and 40 deg) and divide it by the peak displacement value selected from Table 2.

One of the most interesting conclusions from the test results recorded in Table 1 is the large displacement required to reach residual strength values—frequently several centimeters. Data on displacements normal to the shear plane were not presented in Tables 1 and 2. In all laboratory tests where normal displacements were given, shearing displacement was accompanied by dilation (thickening of the joint zone). The peak dilation occurs well after the peak stress at low normal pressures but right at the peak shear stress in tests run under high normal pressure. Several *in situ* block shear tests on clay-filled joints showed slight joint closure early in the shear-displacement curve; however, dilation always occurred eventually.

Deere et al [*10*] proposed a bilinear shear strength envelope for irregular, clean joints, derived from the fact that the failure mechanism at low normal pressures is rising along joint asperities. If the asperities have an inclination α with the mean shear plane, the shear failure envelope for the joint would rise from the origin at an angle $\Phi_{eff} = \Phi + \alpha$. This implies that dilation should occur with the ratio of normal to shear displacement equal to tan α, that is,

$$u_n = u_s \tan (\Phi_{eff} - \Phi) \qquad (1)$$

where:

u_n = displacement normal to the average shear plane,

u_s = displacement in the direction of the average shear plane,

Φ_{eff} = inclination of Mohr envelope at low normal pressures, and

Φ = actual friction angle of the joint surface.

Bray [*11*] showed that a rock mass with multiple joint sets will develop gaps as slip occurs on a single joint set. Thereafter, the gaps will tend either to open further, with $\Phi_{eff} = \Phi + \alpha$, or to close, with $\Phi_{eff} = \Phi - \alpha$, depending upon the direction of applied forces. In the former case, the normal displacement would tend to loosen the rock mass, whereas, in the latter, the rock mass would become more dense as a result of shearing. Eq 1, derived by Bray, applies in either case.

Obviously, irregular joints filled with clay, crushed mica, or other compressible material could also compact, or consolidate, under shear load. As noted above, the only tests in which compacting rather than dilating joint behavior was noted were block shear tests conducted in the field on clay-filled joints.

In Type 4 joints, there has been insufficient experimental work done to define the relative influences on joint strength and deformability of the fundamental variables of the material: joint aperture, joint roughness, degree of filling, and the relative properties of filling and rock material. To illuminate these relationships, an experiment was performed in which irregular joints cast in a plaster of paris-celite mix and filled with finely crushed mica to varying degrees were tested in direct shear. The joint specimens were of 2.50-in. diameter with 4½ teeth having an amplitude of 0.16 in., wavelength of 0.56 in., and inclination α of 30 deg; the tests were carried out in plane strain with constant normal force. Series of tests were completed to determine the properties of plaster alone, mica alone, unfilled joints, and joints filled with mica to a thickness of 0.06, 0.12, and 0.20 in. (measured perpendicularly to the mean plane of the joints).

Figure 10*b* presents the load-deformation data for the test series, Table 4 summarizes the strength-displacement-stiffness relationships, and Fig. 10*c* presents the data on normal displacements during the test. The results may be stated simply (Fig. 11): for the joint tested, the stiffness and strength decreased slowly with the degree of joint filling such that at 125 percent filling[6] the joint stiffness was still twice that of the filling material alone, while the joint strength remained 50 percent greater than the filling strength. This behavior is due to the local concentrations of normal stress in the joint as the asperities approach each other during shear-

[6] Percent joint filling is here defined as (thickness of filling/amplitude of irregularity) × 100.

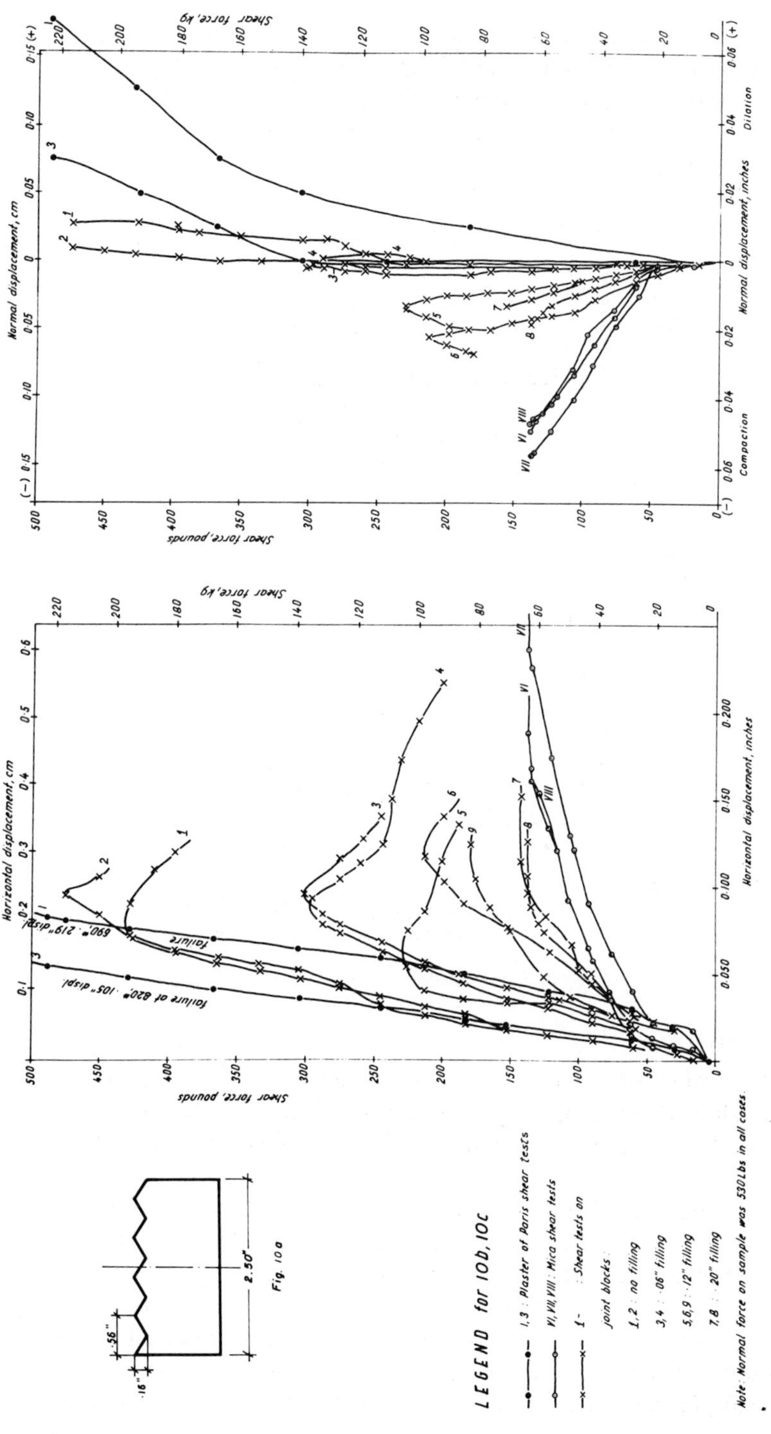

FIG. 10—*Load-deformation curves for direct shear tests on mica-filled joints.*

ing. The experiment indicates that sheared zones and filled joints are likely to have strength and deformability properties appreciably different from those of the filling material itself.

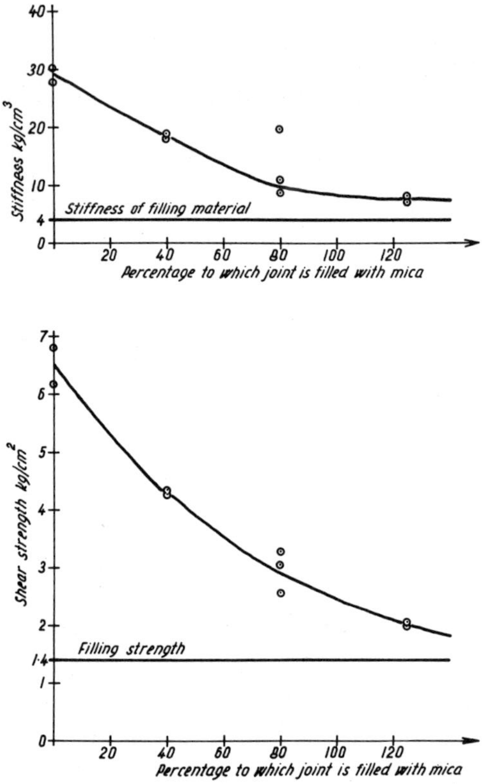

FIG. 11—*Effect of thickness of joint filling on stiffness and strength.*

Conclusions

A number of sampling and testing techniques exist through which the strength and deformability of naturally occurring discontinuities may be assessed. Less attention has been given to displacement and stiffness values in recent literature than to strength, probably because of the predominant use of limit-equilibrium methods in stability analyses of slopes and foundations. Finite element and physical model methods of analysis require deformability as well as strength data. From a review of data from almost 200 tests, it has been possible to group all joints into four types: (1) healed and incipient joints, (2) smooth, clean fractures, (3) rough, clean fractures, and (4) filled joints, sheared zones, and seams. Actual values of peak displacement do not vary appreciably within each group and average values are presented. The tangential stiffness values can be crudely approximated, if the strength is known, by introducing these average displacements. However, almost nothing is known of the correct values for normal stiffness; thus the best policy is to take advantage of the developing art and perform either laboratory tests on undisturbed joint samples or field tests on joints in place, conducted under conditions of loading meaningful for the work in question.

Acknowledgments

I am indebted to the individuals and companies who gave me permission to study their methods and results of testing weakness planes in rock: Pierre Londe and J. Bernaix of Coyne and Bellier; Dusan Krsmanovic and his associates, Messrs. Langof, Tufa, and Dolarovic, at the Faculty of Civil Engineering, Sarajevo; Messrs. Knill and DeFreitas in the Geology Department at Imperial College; Evert Hoek, David Pentz, and John Verge of the Interdepartmental Rock Mechanics Project, Royal School of Mines, Imperial College; Jan Eurenius and Hans Fagerström of Vattenbyggnadsbyrån, Stockholm; Donald Deere and James Coulson of the Department of Civil Engineering, University of Illinois; Klaus John of Teknische Hochschule, Karlsruhe, Germany; Messrs. Bratislav Kujundzic, Brcic, and Marcowitz of the Jaroslav Cerni Institute, Belgrade, Yugoslavia; Karel Drozd of IGHP, Prague, Czechoslovakia; and Manuel Rocha and Ing. Barosso of the National Civil Engineering Laboratory, Lisbon, Portugal. We appreciate the generosity of Electricite de France in permitting reproduction of figures from a Coyne and Bellier report about Vouglans Dam.

I also wish to thank Ashraf Mahtab, Jan Roggeveen, and T. Yasue for help in reviewing the test data and for performing experimental work on the mica-filled joints. The helpful comments of Alain Loriot de Rouvray are also acknowledged.

Finally, I am grateful to the Omaha District of the Corps of Engineers, U. S. Army, and in particular to Carl DiStefano, for support and encouragement in this area of research.

References

[*1*] Krsmanovic, D., *Geotechnique,* GTNGA, Vol. 17, June 1967, pp. 145–160.

[*2*] Krsmanovic, D. and Langof, Z., *Rock Mechanics and Engineering Geology,* Supp. I. (Principles in the Field of Geomechanics), 1964, pp. 20–30.

[*3*] Krsmanovic, D., Tufo, M. and Langof, Z., "Shear Strength of Rock Masses and Possibilities of its Reproduction on Models," *Proceedings,* First Congress of the International Society for Rock Mechanics, Lisbon, Vol. 1, 1966, pp. 537–542.

[*4*] Corps of Engineers, Missouri River Division Lab, "A Study of Rock Joint Movement Under Triaxial Loading," *MRD Laboratory Report No. 65/365,* Sept. 1965.

[*5*] Heuze, F. and Goodman, R. E., chapter 9 in *Status of Practical Rock Me-*

chanics, Proceedings, Ninth Symposium on Rock Mechanics, American Institute of Mining and Metallurgical Engineers, 1968.

[*6*] Jaeger, J. C., "The Frictional Properties of Joints in Rock," *Geofisica pura e Appicata,* GFPAA, Vol. 43, 1959, p. 148.

[*7*] Eurenius, J. and Fagerstrom, H., *Geotechnique,* GTNQA, Vol. 19, No. 1, March 1969, pp. 133–139.

[*8*] Goodman, R. E., Taylor, R. L. and Brekke, T., *Journal of Soil Mechanics and Foundations Division,* Proceedings of the American Society of Civil Engineering, JSFEA, Vol. 94, No. SM3, May 1968, pp. 637–659.

[*9*] Hoek, E. and Pentz, D. L., "The Stability of Open Pit Mines," Imperial College Rock Mechanics Research Report No. 5, Interdepartmental Rock Mechanics Project, Imperial College, London SW 7, England, 1968.

[*10*] Deere, D. U., Hendron, A. J., Patton, F. D. and Cording, E. J., chapter 11 of *Failure and Breakage of Rock; Proceedings,* Eighth Symposium on Rock Mechanics, American Institute of Mining and Metallurgical Engineers, 1967.

[*11*] Bray, J. W., *Rock Mechanics and Engineering Geology,* Vol. 5, Nos. 2 and 3, 1967, pp. 117–136.

Special References — Partial Listing of Unpublished Reports Used as Source Material

[*12*] Bernaix, J., "Etude des Proprietes Mecaniques d'un Echantillon du Gneiss de Malpasset," Ecole Polytechnique, Laboratoire de Mecanique des Solides, 1967.

[*13*] Coyne and Bellier, "Etude de Roches du Fondation Cis aillement sur Carottes, 300-mm Diametre," Rapport pour Barrage de La Rouviere, 1965.

[*14*] Coyne and Bellier, "Etudes du Cisaillement le Long des Joints," Barrage de Vouglans, (Electricite de France), Dossier d'execution, Annexe II, Vol. 3, 1967.

[*15*] Delgado, V. L., "Strength of Joints in Rock Masses," Imperial College, London SW 7, England, 1967.

[*16*] Institute za vodoprivredu Jaroslav Cerni, "Khajuri Kach Dam, Design of Geotechnical Investigations, Phase 3."

[*17*] Bernaix, J., Londe, P., Habib, P. et al, "Essais et Calculs de Mecanique des Roches Appliqués a l étude de la securité des Appuis de'un Barrage-voute-Exemple de Vouglans," *Proceedings*, 9th Congress on Large Dams, Istanbul, Question 32, Report 49, 1967.

[*18*] Drozd, K., "Vysetreni Smykove Pevnosti Pevynch Rospukanych Hornin *In Situ," V E D,* rada HIG, sv. 3, 1965.

[*19*] Drozd, K., "Variations in the Shear Strength of a Rock Mass Depending on the Displacement of the Test Blocks," *Proceedings,* Geotechnical Conference, Vol. 1, Norwegian Geotechnical Institute, Oslo, 1968.

[*20*] Grishin, M. M. and Erdokimov, P. D., "Shear Strength of Structures Built on Rock", *Proceedings,* Fifth International Conference on Soil Mechanics and Foundation Engineering, Paris, 1961, pp. 649–654.

G. B. Wallace[1]

General Discussion

A good *in situ* modulus test is one that will yield a modulus that agrees with measured prototype deformations. There are two basic components of deformation: closure of fractures and joints and elastic strains. Usually joint closure contributes much more to deformation of a rock mass than elastic strains. A good test will have a measuring system that can distinguish between movements that are largely due to joint closure and those principally due to elastic strains. Knowing the component parts of the deformation, a more realistic extrapolation can be made for larger rock masses based on the rock types, fractures, joints and shear zones observed in borehole television surveys and logs of drill holes blanketing the foundation area.

The size and geometry of the test surface should be based on the prototype boundary conditions and the extent of discontinuities in the rock mass. For example, if a dam or pressure tunnel is loading a large volume of rock having significant shear zones, then a large radial test will be best suited for determining the load-time deformation properties of the rock mass. On the other hand, if the rock is homogeneous or contains only tight uniform joint sets, a reliable modulus can be obtained at much less cost by utilizing smaller plate jacking tests. For medium size foundations such as those for heavy buildings, piers, retaining walls, and so forth, plate jacking tests may be sufficient even when significant discontinuities are present. Deformations of foundations for small structures containing reasonably uniform rocks are frequently estimated with sufficient accuracy solely from laboratory tests combined with experience and judgment based on careful inspection of drill cores.

The intensity and cycling of test loads should, of course, be based on prototype loads, and in some cases the deformability of the rock must also be considered. Peak test pressures which are about 50 percent higher than the estimated prototype loads are frequently used. However, for very

[1] Engineer, Structural and Rock Mechanics Section, Division of Research, U. S. Bureau of Reclamation, Denver, Colo.

deformable rock it may be desirable to lower this somewhat to provide a greater accuracy of measurements at the design load and reduce the cost of test equipment. The location and orientation of the test load is also important. To evaluate the influence of joints deep within an abutment for an arch dam, it is usually necessary to drill an exploratory tunnel into the abutment having a test adit located sufficiently close to the joints, that they will be loaded by the test at the prototype intensity and oriented so that test load is parallel to the line of thrust from the dam.

Creep is sometimes very important, especially for pressure tunnels designed to utilize the rock for partial support. If the rock creeps under the working load, its contribution of restraint is reduced accordingly.

An abutment for an arch dam which is extremely inelastic may "set" when unloaded and not maintain full contact with the dam as it resumes its unstrained position. Although a good measure of primary creep and set may be obtained *in situ,* it is probably not often feasible to leave the test load on long enough to completely define the creep equation or determine the ultimate set upon unloading. Under design loads, primary creep is usually much larger than secondary creep. Therefore a practical estimate can be obtained by combining primary creep values obtained from *in situ* tests with secondary creep values based on long-time laboratory tests.

ASTM test methods are used to:

1. Control quality of products
2. Determine design constants required by a code or specification
3. Evaluate different materials for research purposes

At present rock mechanics is not greatly concerned with the first two uses. However, there is considerable effort expended on determining the physical properties of different rock masses. For example, many plate jacking tests are performed by various engineering organizations in a period of a year. If these were performed in accordance with a standard test method, the results over several years could be correlated and reduced to guide lines on the influence of different rock types, fractures, and joint sets of various configurations on the deformation characteristics of rock masses. The cost and availability of test equipment might also be reduced as a result of an acceptable standard test method.